GUIDELINES FOR

PROCESS EQUIPMENT RELIABILITY DATA

WITH DATA TABLES

Publications Available from the
CENTER FOR CHEMICAL PROCESS SAFETY
of the
AMERICAN INSTITUTE OF CHEMICAL ENGINEERS

Guidelines for Technical Management of Chemical Process Safety

Guidelines for Chemical Process Quantitative Risk Analysis

Guidelines for Process Equipment Reliability Data with Data Tables

Guidelines for Vapor Release Mitigation

Guidelines for Safe Storage and Handling of High Toxic Hazard Materials

Guidelines for Use of Vapor Cloud Dispersion Models

Workbook of Test Cases for Vapor Cloud Source Dispersion Models

Guidelines for Hazard Evaluation Procedures

Proceedings of the International Symposium on Runaway Reactions, 1989

Proceedings of the International Conference on Vapor Cloud Modeling, 1987

Proceedings of the International Symposium on Preventing Major Chemical Accidents, 1987

GUIDELINES FOR

PROCESS EQUIPMENT RELIABILITY DATA

WITH DATA TABLES

CENTER FOR CHEMICAL PROCESS SAFETY
of the
American Institute of Chemical Engineers
345 East 47th Street, New York, New York 10017

Library of Congress Cataloging-in-Publication Data

Guidelines for process equipment reliability data
with data tables

Bibliography: p
Includes index.
1. Chemical plants—Equipment and supplies—Reliability. I. American Institute of Chemical Engineers. Center for Chemical Process Safety.
TP155.5.G78 1989 660.2′83 88-36039
ISBN 0-8169-0422-7

This book is available at a special discount when ordered in bulk quantities. For information, contact the Center for Chemical Process Safety at the address shown above.

Contents

Preface

The American Institute of Chemical Engineers (AIChE) has a 30-year history of involvement with process safety and loss control for chemical and petrochemical plants. Through its strong ties with process designers, builders and operators, safety professionals, and academia, the AIChE has enhanced communication and fostered improvement in the high safety standards of the industry. Its publications and symposia have become an information resource for the chemical engineering profession on the causes of incidents and means of prevention.

The Center for Chemical Process Safety (CCPS), a directorate of AIChE, was established in 1985 to intensify development and dissemination of the latest scientific and engineering practices for prevention and mitigation of catastrophic incidents involving hazardous materials. Since its founding, CCPS has co-sponsored several international, technical symposia and has published a number of books. These include four volumes in its *Guidelines* series, the proceedings of three meetings, a technical workbook, and the first in a series of publications on the technical management of chemical process safety. In addition, material has been developed to help integrate process safety into undergraduate chemical engineering programs. CCPS research projects now in progress will yield new data for improved process safety.

Over 50 corporations from all segments of the chemical process industries (CPI) support the Center. They help fund the Center; they select CCPS projects relevant to improved process safety; and they furnish the professionals who give the Center's works technical direction and substance.

The Center for Chemical Process Safety's projects fall into a number of general topic areas that comprise a comprehensive program. These topic areas include identification of hazards and analysis of risks, prevention and mitigation of the hazards identified, and better definition of areas affected by a release of hazardous materials. This book is the latest in the series dealing with hazard identification and risk analysis.

Guidelines for the Use of Vapor Cloud Dispersion Models, the associated *Workbook of Test Cases for Vapor Cloud Source Dispersion Models* and research now in progress are directed toward a more complete understanding of the geographic areas affected by a release to the atmosphere.

Guidelines for Safe Storage and Handling of High Toxic Hazard Materials and *Guidelines for Vapor Release Mitigation* both present engineering practices and operating techniques to prevent and mitigate releases. A new series under development on the fundamentals and systems necessary for successful technical management of process safety is the forerunner of several new projects emphasizing the technologies of prevention and mitigation.

Considerable interest has been generated in hazard identification and risk analysis techniques, which provide a systematic means to help reduce and manage chemical process risks. CCPS has undertaken a series of *Guidelines* covering many aspects of the subjects to provide the latest information and useful techniques for the engineer in the

CPI. The first book, *Guidelines for Hazard Evaluation Procedures* (HEP Guidelines), covers methods for identifying and qualitatively assessing chemical process hazards.

Guidelines for Chemical Process Quantitative Risk Analysis (CPQRA Guidelines) builds on the earlier work to show the engineer how to make quantitative estimates of the risk of the hazards identified. The quantitative estimates can identify the major contributors to risk. They can also help to define the most effective ways to a safer process by indicating relative risk reduction from proposed alternate process safeguards and measures.

This book supplements CPQRA Guidelines by providing information on obtaining some equipment reliability data needed for quantitative risk analyses. It deals, therefore, with rates of equipment failures, the number of equipment failures in 1 million operating hours or in 1000 demands on the equipment. The means to improve equipment performance and data on causes of equipment failure are a segment of reliability engineering and are not addressed here. Human error rates, also needed for a CPQRA, and human reliability in CPI operations will be addressed in another Guidelines presently under development. As discussed in the CPQRA Guidelines, a full risk analysis is not always necessary to fully understand the hazards of a process. However, when one is needed, generic reliability data are often the only data available to a risk analyst. Large plant-specific data bases are seldom available. Because of the many uncertainties inherent in risk analysis techniques, generic data are often sufficient to show major contributors to risk and generate useful results. Helping the reader obtain such generic data is a basic purpose of this book.

The most desirable source of equipment reliability data for a CPQRA is the operating experience of the process and plant being studied. Therefore, a chapter of this book provides information that will help an engineer locate "raw" plant reliability data and convert them to failure rates. However, the quality and confidence level of the plant-specific data may be questionable because of operating and maintenance procedures, short relevant operating experience, and limited pieces of equipment available for study. The best data to use in a CPQRA are often a combination of generic and plant-specific data.

Selection of any equipment reliability data for use in a CPQRA requires good engineering judgment. When using generic failure rate data for a class of equipment in a specified service under a particular operating and maintenance strategy, the engineer or risk analyst must decide if the data are applicable or require modification to compensate for differences in the operating situations. Similarly, engineering judgments are required for data from a specific plant and process where there is usually a limited amount of data available and a high degree of uncertainty about whether the available values are representative. Consequently, another purpose of this book is to present information on failure rates and sources of data that can help the engineer form better engineering judgments about the data to be used. It is important to realize that some situations may require the judgment of an expert.

Making equipment reliability data commonly available requires collection of raw data, conversion of those data into failure rates, and a framework or taxonomy in which the failure rates can be stored. Unless all these tasks are coordinated, there may be no way of fitting them together to produce usable, classified reliability data. In this book, we have attempted to make these three areas, often carried out completely independently, compatible so that any data collected in the future using this book can be easily added to the store of generic reliability data.

The CCPS Taxonomy developed for this book is one step toward accumulating and collating equipment reliability data for the CPI. Ideally, it will be expanded and modified as more companies make chemical process equipment failure rates and reliability data available. We expect that CCPS will update this book and the CCPS generic data base as new information becomes available. The taxonomy may also require modification where experience shows it is needed. We would appreciate any contribution from readers to these ends.

Acknowledgments

GUIDELINES FOR

Process Equipment Reliability Data with Data Tables

Prepared by the

Equipment Reliability Data Subcommittee

of the

CENTER FOR CHEMICAL PROCESS SAFETY

and

SCIENCE APPLICATIONS INTERNATIONAL CORPORATION

The American Institute of Chemical Engineers (AIChE) wishes to thank the Center for Chemical Process Safety (CCPS) and those involved in its operation, including its many sponsors whose funding made this project possible; the members of its Technical Steering Committee who conceived of and supported this Guidelines project; and the members of its Equipment Reliability Data Subcommittee for their dedicated efforts, technical contributions, and the guidance necessary for the preparation of this work.

The chairman of the CCPS Equipment Reliability Data Subcommittee was S. Barry Gibson, E.I. du Pont de Nemours & Co., Inc. The subcommittee members were Harold W Thomas, Air Products and Chemicals, Inc.; William H. Ciolek, Amoco Corporation; Joseph C. Sweeney, ARCO Chemical Company; Brian D. Berkey, Hercules Incorporated; Gary R. Van Sciver, Rohm and Haas Company; and William K. Lutz, Union Carbide Corporation. Thomas W. Carmody and Lester H. Wittenberg of the Center for Chemical Process Safety were responsible for the overall administration and coordination of this project.

AIChE also thanks Joseph R. Fragola, General Manager, and Erin P. Collins, Staff Scientist, of the Advanced Technology Division of Science Applications International Corporation (SAIC) for using their expertise in reliability data handling and data base construction to help organize this Guidelines, provide technical information and reliability data, and prepare this book.

The members of the CCPS Equipment Reliability Data Subcommittee wish to thank their employers for providing time to participate in this project; those sponsors and members of the Technical Steering Committee who reviewed and critiqued this book prior to publication; and those many companies in the chemical processing and allied industries that responded to the Subcommittee's survey of available process equipment reliability data.

Glossary

Active equipment: Denotes physical motion or activity in the performance of the equipment's function, as with rotating machinery.

Aggregation: The statistical combination of several data points to form a single data point and confidence interval.

Alternating mode: Hardware operation that alternates between standby and running, for example, a pump with an installed spare, each of which operates for a comparable amount of time.

Availability: The fraction of calendar time a system is fully operational.

Calendar time: The period between starting date and ending date.

Catastrophic failure: A failure that is both sudden and causes termination of one or more fundamental functions.

Chemical Process Industry: The phrase is used loosely to include facilities that manufacture, handle and use chemicals.

Chemical Process Quantitative Risk Analysis(CPQRA): The numerical evaluation of both incident consequences and probabilities or frequencies and their combination into an overall measure of risk.

Component: An equipment part.

Component boundary: See Equipment boundary.

Computerized Aggregate of Reliability Parameters (CARP): A computer code developed by SAIC to: aggregate data sets into a single generic set; determine uncertainty bounds (5th and 95th percentiles); fit raw data to statistical distributions; and print reports documenting determinations made.

Confidence: A statistical measure of uncertainty.

Confidence bounds or limits: The end points of a confidence interval.

Confidence interval: That portion of a distribution which is expected to contain the mean value a certain percentage of time.

Data base: (1) A repository for equipment reliability information categorized to facilitate data retrieval or (2) tabular lists of multiple data vectors, with little text except that needed to explain the data presentation format.

Data cell: A unique compartment of the taxonomy in which data are stored, defined by specific equipment, service and failure descriptions.

Data elements: The basic items that form a data set or data vector; for example, component name, size, failure mode, mean, 5% confidence level, are each a data element.

Data encoding: The assignment of codes and identifiers to data extracted from plant records so that failure rates may be readily calculated.

Data point: A numerical estimate of equipment reliability as a mean or median value of a statistical distribution of the equipment's failure rate or probability.

Data resource: A data base, report, technical paper, journal article, or conversation that contains reliability data; subdivided into Data Bases, Data Sources, and Risk Analyses in this book.

Data sets: A formal or informal collection of information with a cohesive element that distinguishes this data grouping from others; for example, data from a particular facility, data for a particular time, data for a particular component.

Data source: Descriptive text in a given subject area whose primary purpose is to discuss a reliability or risk topic but that also contains some useful reliability data.

Data vector: Only those data elements and numerical values that are used to specify failure characteristics, for example mean, distribution, failure modes.

Data window: A time frame established for a given data study.

Degraded failure: A failure that is gradual or partial; it does not cease all function but compromises that function. It may lower output below a designated point, raise output above a designated point or result in erratic output. A degraded mode might allow only one mode of operation. If left unattended, the degraded mode may result in a catastrophic failure.

Delphi technique: A polling of experts. The Classical Delphi is a single estimate (for each questionnaire) of a single parameter by a single group. The Hybrid Delphi uses a single estimate of multiple parameters submitted by multiple groups. It allows the incorporation of published or recorded data during the polling process.

Demand: (1) A signal or action that should change the state of a device, or (2) an opportunity to act, and thus, to fail.

Demand spectrum: The total number of demands for the data window experienced by the component population, considering test, interface, failure-related maintenance, and automatic and manual initiation demands.

Error bounds: See Confidence interval.

Error factor: The ratio of the 95th percentile value to the median value of a lognormal distribution.

Equipment: A piece of hardware that can be defined in terms of mechanical, electrical or instrumentation components contained within its boundaries.

Equipment boundary: Demarcation of the equipment defining components included and interfaces with excluded piping, electrical,and instrumentation systems.

Event: An occurrence involving equipment performance or human action, or an occurrence external to the system that causes system upset. In this book, an event is associated with an incident either as the cause or a contributing cause of the incident or as a response to the initiating event.

Event Tree Analysis (ETA): A method for illustrating the intermediate and final outcomes that may arise after the occurrence of a selected initial event.

Exposure, demand-related: The historical number of demands experienced by the equipment population.

Exposure hours: An equipment's operating time in hours.

Exposure, time-related: The historical operating time of the equipment population.

Failure frequency: The number of failures that occur divided by either the total elapsed calendar time during which these events occur or by the total number of demands, as applicable.

Failure mode: A symptom, condition or fashion in which hardware fails. A mode might be identified as a loss of function; premature function (function without demand); an out of tolerance condition; or a simple physical characteristic such as a leak (incipient failure mode) observed during inspection.

Failure Modes and Effects Analysis (FMEA): A hazard identification technique in which all known failure modes of components or features of a system are considered in turn and undesired outcomes are noted.

Failure probability: The probability–a value from zero to one–that a piece of equipment will fail on demand (not to be confused with fractional dead time) or will fail in a given time interval.

Failure rate: The number of failures that occur divided by the total elapsed operating time during which the failures occur or the total number of demands, as applicable.

Failure severity: The degree of functional degradation of equipment usually noted through deficient performance; categorized by the terms "catastrophic," "degraded," and "incipient."

Fault Tree Analysis (FTA): A method for logical development of the many contributing failures that might result in an incident.

Fractional dead time: The mean fraction of time in which a component or system is unable to operate on demand.

Generic data: Data that are typical for a system. Such data will not have been collected for the particular system but will have been collected, estimated, or aggregated from many generally similar systems.

Hazard analysis: The identification of undesired events that lead to the materialization of a hazard, the analysis of the mechanisms by which these undesired events could occur, and, usually, the estimation of the consequences.

Hazard and Operability Study (HAZOP): A technique to identify hazards and problems using a series of guide words to study process deviations.

Historical data: Data recorded from actual past experience.

Human error: Physical and cognitive actions by designers, operators, or managers that may contribute to or result in undesired events.

Incestuous data: Data in two or more data sets that are derived from a common origin and may be inadvertently "double-counted" when aggregated.

Incipient failure: An imperfection in the state or condition of hardware such that a degraded or catastrophic failure can be expected to result if corrective action is not taken.

Isolation: The disablement and tagging-out of appropriate interfacing components prior to initiating maintenance on another component.

Likelihood: A measure of the expected occurrence of an event. This may be expressed as a frequency (e.g., events per year); a probability of occurrence during a time interval (e.g., annual probability); or a conditional probability (e.g., probability of occurrence given that a precursor event has occurred).

Mean: The measure of central tendency of a distribution, often referred to as its arithmetic average.

Median: Midpoint of the failure data distribution.

Nonprocess: Industries that do not comprise the CPI as their primary function but that use comparable or equivalent complex equipment systems to perform their function.

Operating mode: The method of operating equipment. See alternating mode, standby mode, running mode.

Operating time: The amount of time a piece of equipment is in its operating mode.

Passive equipment: Refers to hardware that is not physically actuated in order to perform its function (e.g., piping, valve bodies, pump bodies, and storage tanks).

Plant-specific data: Data that pertain to a unique population of equipment specific to a particular operating plant.

Probabilistic Risk Assessment (PRA): A commonly used term in the nuclear industry to describe the quantitative evaluation of risk.

Probability: The expression for the likelihood of occurrence of an event or an event sequence during an interval of time or the likelihood of the success or failure of an event on test or on demand. By definition probability must be expressed as a number ranging from zero to one.

Process medium: The material processed by the equipment.

Process severity: The indication of the degree of aggressiveness of the process medium on the hardware; aggressiveness would include erosion, stress, corrosion, temperature, blockage, etc. Four categories of severity are used in this book: Clean, General Industry, Moderately Severe, Severe. (See Chapter 2 for further explanation of these categories.)

Raw data: The original records from which reliability data are extracted; the facility records of equipment failure, repair, outage, and exposure hours or demands that require analysis and encoding in order to be placed into data elements.

Reliability: The probability that an item is able to perform a required function under stated conditions for a stated period of time or for a stated demand.

Reliability analysis: The determination of reliability of a process, system, or piece of equipment.

Resource: See Data resource.

Risk: A measure of economic loss or human injury in terms of both the incident likelihood and the magnitude of the loss or injury.

Risk analysis: The development of a quantitative estimate of risk based on engineering evaluation and mathematical techniques for incident consequences or frequencies.

Running mode: Normal hardware operation, for example, an unspared compressor that must operate to run the process.

Safety system: Equipment and/or procedures designed to respond to an initiating event to prevent event propagation.

Sample: An equipment population, its exposure period, and stresses—from which a data set is derived.

Standby mode: Hardware operation that is normally not running but must be ready to run, for example, an emergency diesel generator.

Subsystem: A portion of a system.

System: A collection of equipment considered and usually designated by numeric or naming schemes as a cohesive unit by virtue of the function it performs, the operation it sees, and the conditions for its actuation.

System interaction: Failure in one system that propagates to another.

Taxonomy: A hierarchical organization of data cells, where the items contained in a given level have more equipment reliability characteristics in common with each other than they do with items in any other level.

Taxonomy number: The precise address of a data cell as defined by the classification scheme of the CCPS Taxonomy.

Tolerance: A measure of the uncertainty arising from the physical and the environmental differences between members of differing equipment samples when failure rate data are aggregated to produce a final generic data set.

Uncertainty: A measure of doubt that considers confidence and tolerance.

Unavailability: The fraction of calendar time a system is not fully operational.

Acronyms

ABMA	American Boiler Manufacturers Association
ACRS	Advisory Committee on Reactor Safeguards
AIChE	American Institute of Chemical Engineers
ASME	American Society of Mechanical Engineers
ATV	Swedish Thermal Power Reliability Data System
ATWS	Anticipated Transients Without SCRAM
BEARDS	Baseline Events Analysis Reliability Data System
BNL	Brookhaven National Laboratory
BWR	Boiling Water Reactor
CARP	Computerized Aggregation of Reliability Parameters
CCPS	Center for Chemical Process Safety
CFR	Code of Federal Regulation
CLEF	Computerized Library of Equipment Failures
CMA	Chemical Manufacturers Association
COMPI	TNO's Component Failure Data Bank
COVO	Commission for the Safety of the Population at Large—Netherlands
CPI	Chemical Process Industry
CPQRA	Chemical Process Quantitative Risk Analysis
CREDO	Centralized Reliability Data Organization
DBMS	Data Base Management System
DG	Diesel Generator
DOE	Department of Energy
EPRI	Electric Power Research Institute
ERDS	European Reliability Data System
EEC	European Economic Community
ETA	Event Tree Analysis
EuReDatA	European Reliability Data Association
FIRS	Failure and Inventory Reporting System
FMEA	Failure Modes and Effects Analysis
FRAC	Failure Rate Analysis Code
FSAR	Final Safety Analysis Report
FTA	Fault Tree Analysis
GADS	Generating Availability Data System
GIDEP	Government-Industry Data Exchange Program
GPO	U.S. Government Printing Office
GRS	Gesellschaft fur Reaktorsicherheit
HARIS	Hazards and Reliability Information System
HAZOP	Hazard and Operability Study
HEP	Hazard Evaluation Procedures

HERA	Human Error in Risk Assessment
HRA	Human Reliability Analysis
HTGR	High Temperature Gas Cooled Reactor
ICI	Imperial Chemical Industry
IEEE	The Institute of Electrical and Electronics Engineers
INEL	Idaho National Engineering Laboratory
INPO	Institute of Nuclear Power Operations
IPRDS	In-Plant Reliability Data System
IRRAS	Integrated Risk and Reliability Analysis System
ISBN	International Standard Book Number
LER	Licensee Event Report
LMFBR	Liquid Metal Fast Breeder Reactor
LNG	Liquefied Natural Gas
LOCA	Loss of Cooling Accident
LOSP	Loss of Off Site Power
LPG	Liquefied Petroleum Gas
LWR	Light Water Reactor
MOV	Motor Operated Values
MTBF	Mean Time Between Failures
MTBR	Mean Time Between Repair
MTBM	Mean Time Between Maintenance Actions
MTBS	Mean Time Between Shutdowns
NERC	North American Electric Reliability Council
NPAR	Nuclear Plant Aging Research
NPE	Nuclear Power Experience
NPP	Nuclear Power Plant
NPRDS	Nuclear Plant Reliability Data System (sponsored by INPO)
NRC	Nuclear Regulatory Commission
NREP	National Reliability Evaluation Program
NRR	USNRC Office of Nuclear Reactor Regulation
NSAC	Nuclear Safety Analysis Center
NSIC	Nuclear Safety Information Center
NSSS	Nuclear Steam System Supplier
NTIS	National Technical Information Service
NUREG	Document sponsored by NRC
OREDA	Offshore Reliability Data
ORNL	Oak Ridge National Laboratories
PDU	Process Development Unit
PERD	Process Equipment Reliability Data
PRA	Probabilistic Risk Assessment
PWR	Pressurized Water Reactor
QRA	Quantitative Risk Analysis
RAC	Reliability Analysis Center at RADC
RADC	Rome Air Development Center
RCP	Reactor Coolant Pump
RWE	Rheinische Westalisches Elekrizitatswerke
SAIC	Science Applications International Corporation
SNL	Sandia National Laboratories

SRS	Systems Reliability Service, U.K.A.E.A.
SYREL	Systems Reliability Service Data Bank
TNO	Netherlands Organization for Applied Scientific Research
TUV	German Institute for Reactor Safety of the Technical Inspection Association
UKAEA	United Kingdom Atomic Energy Authority
USNRC	United States Nuclear Regulatory Commission
WASH-1400	Reactor Safety Study: An Assessment of Accident Risk in U.S. Commercial Nuclear Power Plants (Source 4.8-9)

1
Introduction

This chapter introduces the need for process equipment failure rate data, defines the scope and organization of this book and the data it contains, and explains how to the use the book.

1.1 Background

The Chemical Process Industry (CPI) uses various quantitative and qualitative techniques to assess the reliability and risk of process equipment, process systems, and chemical manufacturing operations. These techniques identify the interactions of equipment, systems, and persons that have potentially undesirable consequences. In the case of reliability analyses, the undesirable consequences (e.g., plant shutdown, excessive downtime, or production of off-specification product) are those incidents which reduce system profitability through loss of production and increased maintenance costs. In the case of risk analyses, the primary concerns are human injuries, environmental impacts, and system damage caused by occurrence of fires, explosions, toxic material releases, and related hazards. Quantification of risk in terms of the severity of the consequences and the likelihood of occurrence provides the manager of the system with an important decision-making tool. By using the results of a quantitative risk analysis, we are better able to answer such questions as, "Which of several candidate systems poses the least risk?" "Are risk reduction modifications necessary?" and "What modifications would be most effective in reducing risk?"

In performing such a risk analysis, the risk analyst first uses hazard identification techniques such as those presented in *Guidelines for Hazard Evaluation Procedures* (henceforth, HEP Guidelines) such as Failure Modes and Effects Analysis (FMEA) or Hazard and Operability Studies (HAZOP) to identify the incidents or combinations of incidents that must occur to create a given consequence. The analyst may also use techniques such as fault tree analysis or event tree analysis to further define the particular process or equipment failures that will result in the incidents of concern. Using procedures contained in *Guidelines for Chemical Process Quantitative Risk Analysis* (henceforth, CPQRA Guidelines), the severity of the consequences is then evaluated by considering the environment in which the incident occurs and, if necessary, applying techniques such as dispersion, blast, or heat radiation modeling. To evaluate the likelihood of occurrence of the incident, the analyst must know how frequently the contributory failure incidents are likely to occur. Consequently, failure rate data for the equipment involved in the incidents is essential to the risk analysis.

1.2 Guidelines Purpose, Scope, and Organization

1.2.1 Purpose

The primary purpose of this book is to provide the engineer and risk analyst with failure rate data needed to perform a CPQRA. Consequently, the book contains easily accessible data in the CCPS Generic Failure Rate Data Base, information on several available generic data resources, and procedures to develop failure rate data using information from the plant and process being studied. Another purpose is to present an approach that coordinates the collection of raw plant data, their conversion into plant-specific failure data, and their storage using a CPI-oriented taxonomy. This approach will allow future data generated by chemical process facilities to be added to the CCPS Generic Failure Rate Data Base. The book provides specifications for the transfer of data. It is hoped this approach and standardization will stimulate the chemical processing industry to generate and transfer failure rate data to CCPS for industry use. It is also expected that this book and the CCPS Taxonomy will be revised and updated when sufficient new data become available. Finally, this *Guidelines* is written to help engineers and analysts develop an understanding of the derivation, usefulness, and limitation of failure rate data so they can form better judgments about the use of data.

1.2.2 Scope

The data presented in this book are characterized as equipment failures per 10^6 operating hours for time-related failure rates and failures per 10^3 demands for demand-related failure rates. These rates are given for some common CPI equipment. Equipment used solely to transport chemicals is not covered in this book. The cause of equipment failures, the means to improve reliability and the "most" reliable equipment are not addressed. Other types of failure rate data, such as predicted values or estimated values using expert opinion or the Delphi technique, are addressed in the CPQRA Guidelines. Sources of common cause/mode failure data are not addressed. Human error rates, though necessary for CPQRAs, and human performance in CPI facilities will be addressed in a forthcoming Guidelines. Figure 1.1 illustrates the scope of this book in relation to the CPQRA Guidelines. In preparing this book, the CCPS Subcommittee tried to review all published sources of available generic equipment reliability and failure rate data, including reliability studies, published research works, reliability data banks, or government reports that contained information gathered from chemical process, nuclear, offshore oil, and fossil fuel industries around the world. An industry survey was conducted to solicit unpublished data.

1.2.3 Organization

The sections of this book and their contents are:

- **Chapter 1—Introduction:** Discusses the need for process equipment failure rate data, describes the purpose, scope and organization of this book, and explains how to the use it.
- **Chapter 2—Origin, Use, and Limitations of Failure Rate Data:** Explains the meaning of generic and plant-specific data, the difference between time-related and demand-

related failures, issues of confidence and tolerance, what is captured as an equipment failure, the failure model used and the role of the taxonomy.

- **Chapter 3—CCPS Taxonomy:** Explains the CCPS taxonomy. Discusses the rationale and process for its development and the factors considered in its construction.
- **Chapter 4—Data Bases, Sources, and Studies:** Summarizes and characterizes several generic data resources available to risk analysts and process engineers in the CPI. It includes a discussion of the resource search and selection process and the presentation format for the information on resources.
- **Chapter 5—CCPS Generic Failure Rate Data Base:** Contains tables of generic process equipment reliability data that are structured by the CCPS Taxonomy. The data are extracted from data resources in Chapter 4. The chapter includes a discussion of the selection, treatment, and presentation of the data in the Tables.
- **Chapter 6—Collection and Conversion of Plant-Specific Data:** Describes the type of data required and their treatment to develop a plant-specific data set suitable for use or aggregation with other data.
- **Chapter 7—Failure Rate Data Transfer:** Provides a form to facilitate the transfer of plant-specific data to the CCPS Data Base or to combine it with other generic data.
- **Chapter 8—Supplemental References:** A collection of references that describe data collection, analysis, and application techniques but, in general, do not contain reliability data.
- **Appendix A—CCPS Taxonomy:** The full CCPS Taxonomy for process equipment failure rate data.
- **Appendix B—Equipment Index:** Allows the user to determine the taxonomy location for equipment types familiar to the CPI.
- **Appendix C—Matrix of Data Elements in Each Data Resource:** Presents the user with a more detailed summary of the data elements available from each data resource in Chapter 4.
- **Appendix D—Unreviewed Data Bases, Data Sources, and Studies:** Provides a list of data resources that were uncovered too late for review.

1.3 Use of This Guidelines

It is recommended as a first step that the user of the book review the entire volume to become familiar with the various aspects of equipment failure rates that are presented. This can provide a better understanding of the derivation, value, and limitations of generic data. Beyond this, the volume is structured to assist the reader in one or more of three basic tasks. These tasks are:

- locating generic data for use in a CPQRA;
- finding potential data resources for additional data or more information; and
- developing a system for collecting and recording in-plant reliability data.

To find generic data in this book for use in a CPQRA, the reader should first locate the taxonomy number for the equipment under study by referring to Appendix B, Equipment Index. This index shows the taxonomy number for various types of commonly used equipment. Knowing the taxonomy number, the reader can consult the Index of Filled Data Cells (Table 5.2) to determine if the data exist in Chapter 5. Alternatively, the user

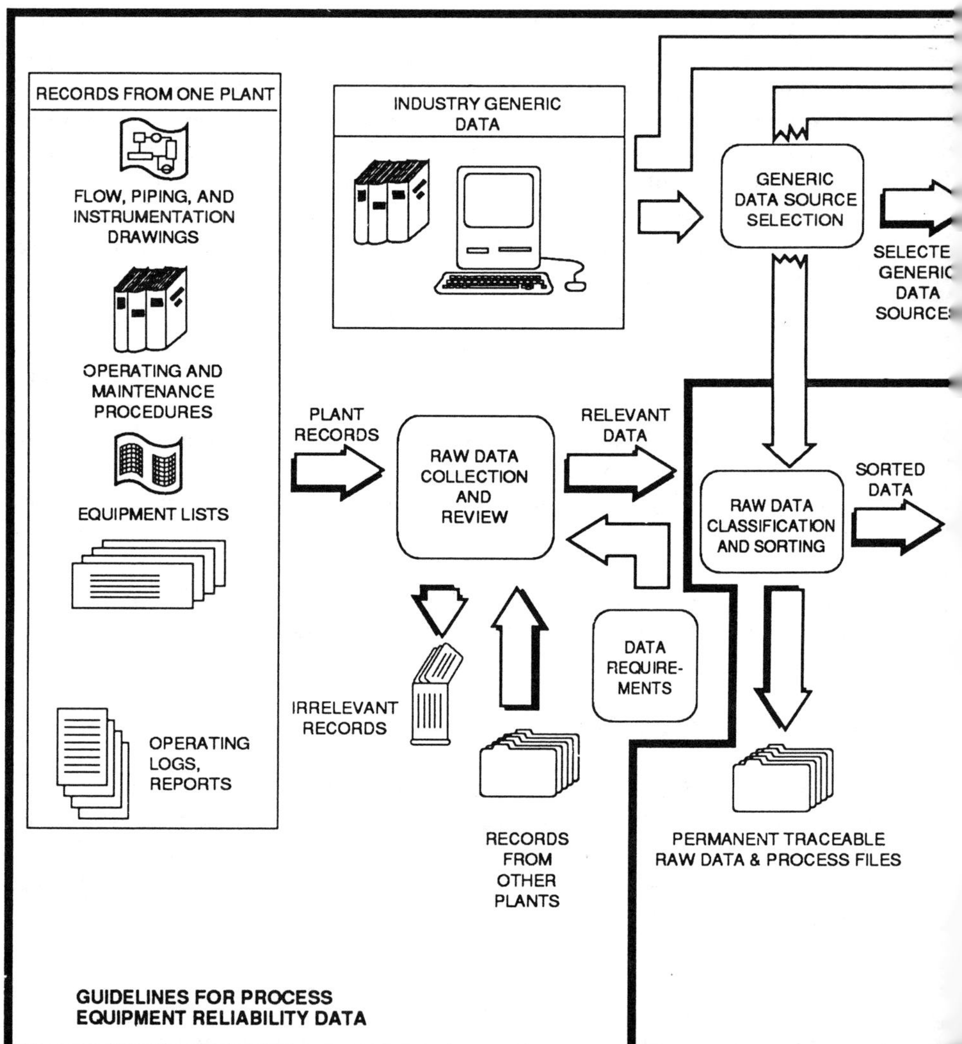

Figure 1.1. *Process equipment reliability data: data sources, data flow, and data use.*

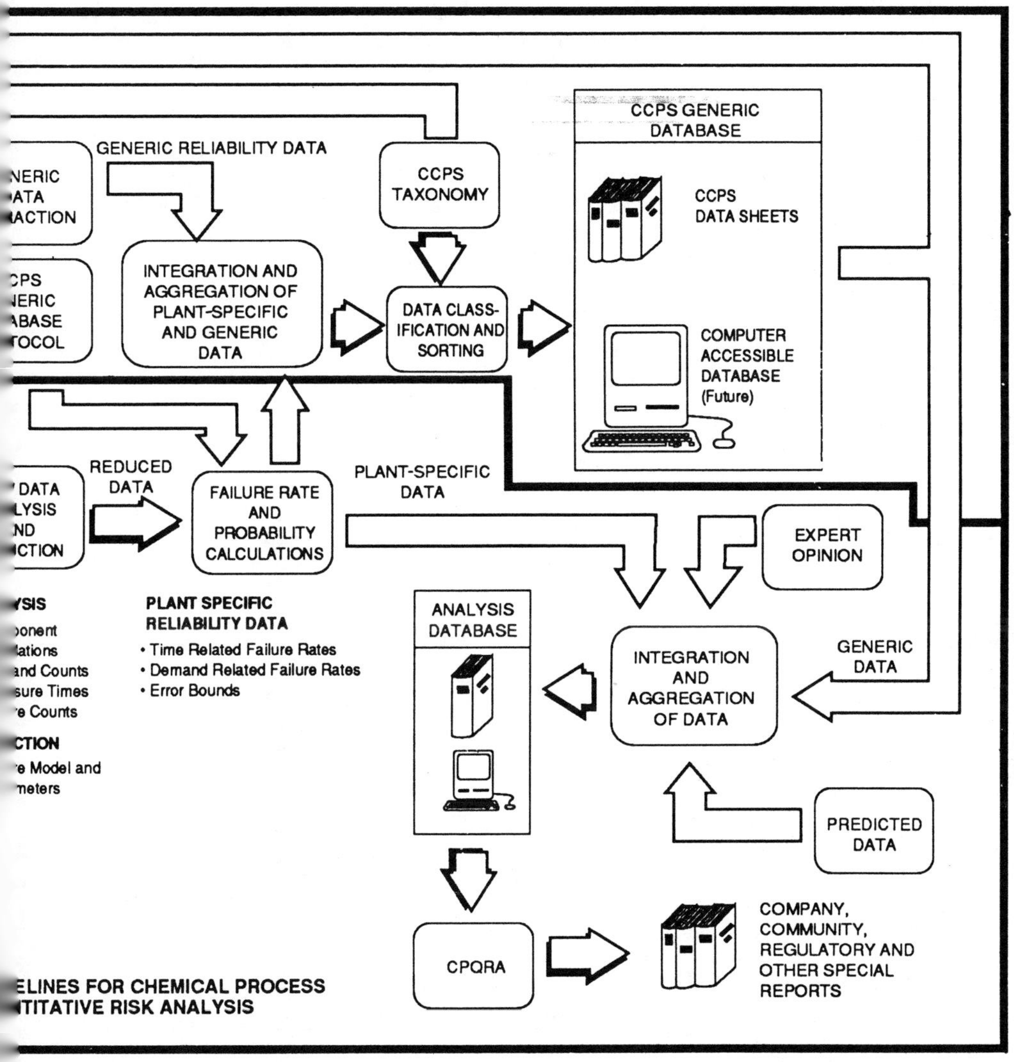

GENERIC RELIABILITY DATA
NERIC
ATA
ACTION
CPS
NERIC
BASE
TOCOL
CCPS TAXONOMY
INTEGRATION AND AGGREGATION OF PLANT-SPECIFIC AND GENERIC DATA
DATA CLASS-IFICATION AND SORTING
CCPS GENERIC DATABASE
CCPS DATA SHEETS
COMPUTER ACCESSIBLE DATABASE (Future)
DATA
LYSIS
ND
CTION
REDUCED DATA
FAILURE RATE AND PROBABILITY CALCULATIONS
PLANT-SPECIFIC DATA
EXPERT OPINION
YSIS
ponent
lations
and Counts
sure Times
e Counts
CTION
e Model and
meters
PLANT SPECIFIC RELIABILITY DATA
• Time Related Failure Rates
• Demand Related Failure Rates
• Error Bounds
ANALYSIS DATABASE
INTEGRATION AND AGGREGATION OF DATA
GENERIC DATA
PREDICTED DATA
CPQRA
COMPANY, COMMUNITY, REGULATORY AND OTHER SPECIAL REPORTS
ELINES FOR CHEMICAL PROCESS
TITATIVE RISK ANALYSIS

may look in the Chapter 5 Data Tables once the taxonomy number has been located. Readers who require additional sources of data should refer to the indexes at the beginning of each resource section in Chapter 4 (Sections 4.3, 4.4, 4.5, 4.6, 4.7, or 4.8). These indexes help the reader identify the most useful resources in terms of the type and extent of data presented. Appendix C is also useful, as it provides additional detail about these data resources.

When plant-specific data are required, Chapter 6 discusses how to collect and treat the data so that the resulting failure rates can be used in a CPQRA or be combined with the data in the CCPS Generic Failure Rate Data Base. Chapter 7 provides a form that can be used to transfer these data to CCPS's Generic Failure Rate Data Base.

2
Equipment Failure Rate Data

To properly use failure rate data, the engineer or risk analyst must have an understanding of failure rates, their origin and limitations. This chapter discusses the types and source of failure rate data, the failure model used in computations, the confidence, tolerance and uncertainties in the development of failure rates and taxonomies which can store the data and influence their derivation.

2.1 Sources and Types of Failure Rates

Failure rate data generated from collecting information on equipment failure experience at a plant are referred to as plant-specific data. A characteristic of plant-specific data is that they reflect the plant's process, environment, maintenance practices, and choice and operation of equipment. Data accumulated and aggregated from a variety of plants and industries, such as nuclear power plants, CPI or offshore petroleum platforms, and are called generic data. With inputs from many sources, generic failure rate data can provide a much larger pool of data. However, generic data are derived from equipment of many manufacturers, a number of processes, and many plants with various operating strategies. Consequently, they are much less specific and detailed.

Both of the sources above contain two types of failure rate data used in CPQRAs: time-related failure rates and demand-related failure rates. Time-related failure rates, presented as failures per 10^6 hours, are for equipment that is normally functioning, for example, a running pump, or a temperature transmitter. Data are collected to reflect the number of equipment failures per operating hour or per calendar hour.

Demand-related failure rates are presented as failures per 10^3 demands and are for equipment that is normally static but is called upon to operate at indeterminate intervals, for example, a switch or standby generator. In this case, data are gathered that can be converted to reflect the number of failures per demand on the equipment.

Both time-related failure rates and demand-related failure rates can apply to and be reported for many pieces of equipment. Both types of rates are included in some of the data tables in Chapter 5. If a piece of equipment is in continuous service, such as a transformer, the failure rate is dominated by time-related stresses compared to demand-related stresses. Other failure rates may be dominated by demands. Take a piece of wire and repeatedly bend it. With each bend its probability of catastrophic failure increases. In a relatively short time, if the bending is continued, the wire will fail. On the other hand, the same wire could be installed in a manner that would prevent mechanical bending demands. In this case, the occurrence of catastrophic wire breakage would be remote. In the first instance, the failure rate is dominated by demand stresses and in the second by time-related stresses, such as corrosion.

Another example is a safety valve in standby service. If demands occur very infrequently, time-related stresses such as external corrosion may have a significant influence. Repeated demands in very dirty service could easily lead to faster degradation and failure, whereas repeated demands in lubricated service might actually enhance performance if the failure mode of interest is failure to open. Failure data based on time or demands can also be skewed if the relief valve is initially damaged or installed incorrectly.

The above discussion leads to the conclusion that time-related and demand-related failures for a piece of equipment cannot be equated through a general mathematical relationship. These issues are better dealt with in a data base taxonomy (classification scheme) for equipment reliability data by defining a unique application through equipment description, service description, and failure description.

2.2 Failure Model

A uniform definition of a failure and a method of classifying failures is essential if data from different sources are to be compared. The anatomy of a failure includes the initiating or root cause of a failure that is propagated by contributory causes and results in a failure mode—the effect by which a failure occurs or is observed. Modes include failure to operate, no output, failure to alarm on demand. The end result of a failure sequence is the failure effect, such as no fluid is pumped to the absorber, or a tank overflows. As discussed in Appendix A of IEEE Std. 500-1984,[1] only the equipment failure mode is relevant for data that are needed in a CPQRA. The failure model used in this book is based upon those in the IEEE publication and IPRDS.[2]

Failures can occur in two general types of equipment—active and passive—explained as follows:

- **Active**: Physical motion or activity in the performance of an equipment's function (e.g., rotating equipment).

- **Passive**: Equipment not physically actuated in order to perform its function (e.g., piping, storage tanks).

Failure modes vary in degree of magnitude, for example, a pump may have no output or may have its output restricted. Consequently, failure modes have been divided into three categories of "severity," which are defined as follows:

- **Catastrophic:** A failure that is both sudden and causes termination of one or more fundamental functions.
- **Degraded:** A failure that is gradual or partial.
- **Incipient:** An imperfection in the state or condition of equipment such that a degraded or catastrophic failure can be expected to result if corrective action is not taken.

There are a number of failure modes for the three failure severities and for active and passive equipment. Figures 2.1 and 2.2 illustrate these failure modes and severities by type of equipment.

			Failure Severity		
			Catastrophic	**Degraded**	**Incipient**
Change in item or equipment condition	Change in operation		1. Failure to operate (run) 2. No output	1. Low output 2. High output 3. Erratic output 4. Locked in one mode of operation 5. Output above or below specified requirements	Discovered through: 1. Local inspection (overheating, leaks, contamination, noise, severe vibration, odor, cracks, etc) 2. Testing: (output above or below specified limits while in stand-by mode of operation) 3. Monitoring (trend towards failure)
	Change of state	Change without demand	A spurious: 1. Start/Stop 2. Insertion 3. Withdrawal 4. Actuation 5. Response 6. Opening 7. Closing	1. Premature or delayed actuation (an actuation that occurs out of timing sequence) 2. Won't stay open or closed	Discovered through: 1. Testing: Failure or diminished ability to transmit or retain energy during the stand-by mode of operation 2. Local inspection
		No change on demand	Failure to: 1. Start 2. Stop 3. Insert 4. Withdraw 5. Actuate 6. Respond to command 7. Open 8. Close	Improper Response: 1. Partially open, close, etc 2. Oscillation (failure to assume a fixed position)	

Figure 2.1 *Active equipment failure modes. Reprinted from ANSI/IEEE Std. 500-1984, © 1984 by the IEEE, with permission of the IEEE Standards Department.*

2.3 Taxonomy

In order to have readily accessible and retrievable failure rate data that can be compared and aggregated with similar data, a logic to classify and store the data is needed. Such a logic and structure, called a taxonomy, is based on equipment and process characteristics affecting reliability, and it creates categories of equipment having similar failure rates. There are many ways to classify equipment failure systematically and consequently structure a taxonomy. Some of the characteristics that can define these categories are the equipment, its function, size, speed, operating mode, and failure mode. A taxonomy's hierarchy, based on these characteristics, creates a multitude of data cells, each with a unique address to house failure rate data for each specifically defined piece of process equipment and its service. The classification scheme developed by CCPS is presented in Chapter 3.

<table>
<tr><th></th><th colspan="4">Failure Severity</th></tr>
<tr><th></th><th>Catastrophic</th><th>Degraded</th><th colspan="2">Incipient</th></tr>
<tr>
<td rowspan="2">Change in item or equipment condition</td>
<td rowspan="2">Failure to retain or transmit energy

1.0 Breach of pressure or static fluid boundary
1.1 Major leaks
1.1.1 External leaks
1.1.2 Internal leaks
1.2 Explosions
1.3 Implosions
2.0 Loss of energy transport or exchange capability
2.1 Blocked or stopped flow
2.2 Loss of heat transfer capability (scale buildup)
2.3 Major heat loss (loss of insulation)
3.0 Loss of structural integrity
3.1 Failure to support or brace
3.1.1 Fracture (of all members)
3.1.2 Physical distortion (permanent set)
3.1.3 Distortion under load (without perm. set)
3.2 Failure to fasten or join
3.2.1 Removable fastener failure
3.2.2 Failure of permanent joint
3.2.2.1 Weld failure
3.2.2.2 Imbed failure</td>
<td rowspan="2">Diminished ability to retain or transmit energy

1.0 Degradation of pressure or static fluid boundary
1.1 Minor leaks
1.1.1 External leaks
1.1.2 Internal leaks
2.0 Interference with energy transport or exchange capability
2.1 Restricted flow
2.2 Reduced heat transfer capability
2.3 Minor heat loss
3.0 Structural integrity compromised
3.1 Reduced support capability
3.1.1 Fracture of part of the structural members
3.1.2 Minor physical distortion
3.2 Partial failure to fasten or join</td>
<td>Change in operation</td>
<td>(1) Testing: Failure of diminished ability to transmit or retain energy during the energized mode of operation
(2) Local inspection (leaks, vibration, odor, cracks, etc)
(3) Monitoring: Monitoring trend towards failure, during the energized mode of operation</td>
</tr>
<tr>
<td>Change of state</td>
<td>(1) Testing: Failure or diminished ability to transmit or retain energy during the stand-by mode of operation
(2) Local inspection</td>
</tr>
</table>

Figure 2.2 *Passive equipment failure modes. Reprinted from ANSI/IEEE Std. 500-1984, © 1984 by the IEEE, with permission of the IEEE Standards Department.*

A taxonomy serves as the receptor of all failure rate data and the model for new data. The descriptors of each data cell govern the collection and conversion of raw plant data and placement of the plant-specific failure rate data into the taxonomy. Having well-defined and generally accepted descriptors for each data cell enhances gathering and combining generic data by ensuring that "apples are compared to apples" when data are aggregated from different resources. Understanding the taxonomy structure provides insight into the characterization of equipment failure rates.

2.4 Confidence and Tolerance

Failure rates are computed by dividing the total number of failures for the equipment population under study by the equipment's total exposure hours (for time-related rates) or by the total demands upon the equipment (for demand-related rates). In plant operations, there are a large number of unmeasured and varying influences on both numerator and denominator throughout the study period or during data processing. Accordingly, a statistical approach is necessary to develop failure rates that represent the true values.

Equipment failure rate data points carry varying degrees of uncertainty expressed by two measures, confidence and tolerance. Confidence, the statistical measurement of uncertainty, expresses how well the experimentally measured parameter represents the actual parameter. Confidence in the data increases as the sample size is increased.

Tolerance uncertainty arises from the physical and the environmental differences among members of differing equipment samples when failure rate data are aggregated to produce a final generic data set. Increasing the number of sources used to obtain the final data set will most likely increase the tolerance uncertainty.

The ideal situation when performing a CPQRA is to have sufficient plant-specific data for each piece of equipment. However, there are many variables in the process, maintenance practices, and data collection that can fluctuate throughout a study period and have a major influence on the results: intensified preventive maintenance can lower and eliminate failures; changes in process conditions may severely exacerbate fouling tendencies or corrosion rates; equipment may be upgraded and even replaced during the study extending operating life; many failures may have been missed; many failures may be wrongly recorded such as a reported pump failure when the push button was really at fault. Since populations and operating time are limited in most plant studies and the number of failures may be heavily biased by the variations noted, plant-specific failure rates may carry little statistical confidence.

Confidence that the calculated failure rate is a good estimate of the true rate can be increased by lengthening the study or sample time. Adding another population of the same equipment under the identical circumstances to the original population will reduce uncertainties and increase confidence in the calculated failure rates.

Plant-specific data are frequently unavailable or are low in their level of confidence. Further, this source of data cannot provide information on equipment not in use at the plant, nor can it do more than suggest how plant equipment might behave under different circumstances. Since data collection is very difficult, using shared or generic data is one way of resolving these problems without the expense of extensive data collection systems.

Frequently, the only way to gather sufficient data for a CPQRA is to build a data set using inputs from other plants within the company or from other available resources. Generic data provide less specific and detailed data, but can draw upon a much larger equipment population, representing more exposure time, and present a much more realistic range of data than that limited to a single plant.

However, the data that are contributed to a generic failure rate data base are rarely for identical equipment and may represent many different circumstances. Generic data must be chosen carefully because aggregating generic and plant-specific data may not improve the statistical uncertainty associated with the final data point, owing to change in tolerance.

2.5 Sources of Variation in Failure Rates

A failure rate generated from collecting data on a system will be dependent upon all the circumstances under which the system operates. Consequently, the failure rate data should only be used for predictions on a system in which the circumstances are identical. Otherwise, the failure rate applicable to the second system will need to be adjusted.

Unfortunately, the circumstances of a data collection exercise are rarely adequately described; and therefore, any data will be based on some explicit assumptions, some implicit assumptions, and some assumptions that are completely ignored.

It is important to appreciate that a failure rate is not an intrinsic and immutable property of a piece of equipment, and an engineer involved either in collecting or using data must fully understand the factors that influence failure rate derivation and use. This section discusses many of the circumstances that can create variations in failure rates.

2.5.1 *Equipment Boundary*

The various data cells in a taxonomy include a written description of the equipment and a boundary diagram to identify exactly what equipment is included within the cell. Any change in the boundary diagram or deviation from it in failure attribution during data processing will influence the failure rate and its comparability with others.

2.5.2 *Taxonomy Level Breakdown*

The various levels of the taxonomy represent factors that have an impact on failure rate. For example, lined pipe (CCPS taxonomy number 3.2.2) has a level that groups pipe into 0–6" size and over 6". Unless the pipe size is specified, there is no way of knowing whether a given failure rate came from the 0–6" or the over 6" range.

2.5.3 *Process Severity*

In the CCPS Taxonomy, four degrees of severity, from "clean" to "severe," are used to characterize the process medium—the material being handled by the equipment—and its influence on reliability. In some cases, the severity will be unknown. Even if a severity is listed, doubt may exist about its value, since the definitions of severity are fairly subjective.

2.5.4 *Environment*

Another influence on equipment reliability is the environment/application of the equipment. A component working on a rocket into space is quite likely to have a different failure rate from the same component operating in a plant control room. Such things as external temperature, humidity, vibration, external corrosion, and any other external conditions imposed on the system need to be considered by the engineer or analyst.

2.5.5 *Suitability for Service*

In Table 3.2, a number of factors are listed that were not used as separate levels in the CCPS taxonomy because of assumptions made by the CCPS Subcommittee. The analyst must, wherever possible, try to assess the validity of these assumptions for the particular situation and establish if the equipment represented by the data:

- was properly fabricated;
- used appropriate materials of construction;
- was properly maintained;
- was operated within design conditions;
- was designed to appropriate standards;
- was being used beyond its capabilities.

2.5.6 *Maintenance*

The maintenance strategy for a system will significantly affect both the number and severity of failures:

- An inadequate maintenance program will prevent no failures.
- A cursory routine inspection program will detect some potential failures; for example, low oil level, which could eventually lead to a seizure.
- A full preventive maintenance program will pick up potential failures as incipient failures rather than delaying until they become catastrophic.

2.5.7 *Data Capture*

What should be recorded as a failure is very subjective. For example, low oil level may be considered too trivial to record, and yet it is an incipient failure on the way to a lubrication failure and ultimately equipment seizure. A truck backing into a pump would certainly stop the pump from functioning, but has it been included in the data collected?

Sections 2.5.1 to 2.5.7 are not an exhaustive list of equipment, process, maintenance, and data processing factors that influence failure rate, but they are an attempt to make the reader aware of the problems. The following example illustrates some of the points made:

Consider how different systems might treat the following pump failures:

1. Seal wore out causing a leak
2. Truck backed into pump shattering case
3. Pumping against closed head and overheated
4. Foreign matter in pumped fluid chewed up seal
5. Wet product corroded impeller
6. Suction blocked by foreign body

All of the above events would cause a pump "failure" over a period of time. Therefore, the events would qualify for inclusion in the failure rate. So, at one extreme there might be six catastrophic failures per sample time. However, a data analyst may decide that No. 2 is not a relevant failure since the cause was neither a function of the equipment nor the operational application, but was a mistake by an outside agent. The same might be said of No. 3.

If a plant had periodic inspections, the impeller corrosion in No. 5 might be detected before it became a significant problem, thereby altering the failure mode from catastrophic to a degraded or an incipient failure. In a plant with routine maintenance, it is possible that Nos. 1 and 5 may be eliminated completely by routine seal and impeller changes.

It is easy to see, therefore, that in one operating system six catastrophic failures would be recorded whereas in others they would range through a combination of catastrophic, degraded, or incipient failures until, with better filters, better operator, frequent scheduled maintenance, all the failures would be eliminated.

2.6 Time-Related and Demand-Related Failure Causes

Although failures are recorded as time related or demand related, the distinction between the two failure types is not always clean cut. The total failures on any piece of equipment are usually a combination of some that are time dependent and some that are demand related. In other words, a piece of equipment that fails on demand may already be in a failed state when the demand arrives, or the demand may actually cause the failure.

On normally operating equipment, because the demand is continuous, all failures can be considered as functions of time; but for equipment that is operated intermittently, the relative proportion of demand- and time-related failures in combination with the frequency of demands and the type of maintenance and inspection program will have an impact on its total failure rate. This point is best illustrated by an example:

Consider the failure of an unloading hose that leaks on being coupled up. For simplicity, it is assumed that the failures are caused by: (1) atmospheric corrosion while the unused hose is waiting for the next unloading or (2) damage from the act of coupling.

The first is clearly a time-related failure and the second a demand-related failure.

A data collection system, or an analyst wanting data, would simply record the number of failures (the numerator) and the number of demands during the period of interest (the denominator). Dividing the numerator by the number of times coupling had occurred would produce the failure rate in failures per demand.

To illustrate the possible variation in failure rates under different operating requirements and maintenance programs assume the following:

- Corrosion-caused failures occur at a rate of two per year.
- Corrosion problems are eliminated by scheduled replacement of the coupling every 3 months.
- Coupling seal damage is caused at the rate of one every 100 couplings made (10 per 10^3 demands).

Consider four cases:

1. A system with no scheduled coupling replacement and 12 unloadings per year.
2. A system with no scheduled coupling replacement and 365 unloadings per year.
3. A system with the coupling replaced every 3 months and 12 unloadings per year.
4. A system with the coupling replaced every 3 months and 365 unloadings per year.

TABLE 2.1

Case	Scheduled Replacement	Unloads per year	No. of Failures Due To Corrosion (1)	Damage (2)	Total (3)=(1)+(2)	No. of Demands (4)	Failures per 10^3 Demand (3)÷(4)×10^3
1	No	12	2	0.12	2.12	12	177
2	No	365	2	3.65	5.65	365	15
3	Yes	12	0	0.12	.12	12	10
4	Yes	365	0	3.65	3.65	365	10

Table 2.1 compares the expected results of data collection for the four cases for a 1-year period.

As illustrated by the above examples, details of how failures have occurred are needed to fully understand the anatomy of a failure and to distinguish between time-related and demand-related failure rates. Defining and classifying failures is not a trivial task, and, because it is often ignored, the final data must be treated with appropriate caution.

2.7 Using Failure Rate Data

When using failure rate data for a CPQRA, the ideal situation is to have valid historical data from the identical equipment in the same application. In most cases, plant-specific data are unavailable or may carry a level of confidence that is too low to allow those data to be used without corroborating data. Risk analysts often overcome these problems by using generic failure rate data as surrogates for or supplements to plant-specific data. Because of the uncertainties inherent in risk analysis methodology, generic failure rate data are frequently adequate to identify the major risk contributors in a process or plant.

Selecting appropriate generic data requires understanding and judgment. In many cases, the analyst can find a number of generic data points that might be used for a CPQRA. Data points chosen for use must provide the level of confidence necessary without creating an unacceptable tolerance uncertainty. The uncertainties of data selection can be reduced by learning as much as possible about data sets, including the taxonomy and equipment boundaries used; the type, design, and construction of the equipment; the process medium; plant operation and maintenance programs; and failure modes. OREDA,[3] IEEE Std. 500-1984[1] and Reliability Data Book for Components in Swedish Nuclear Power Plants[4] are examples of data sets that provide details of taxonomy, data origin, treatment, and limitations. By knowing the background of the data pool, an engineer can more easily choose appropriate data points.

After data have been selected and combined with other generic data or plant-specific data to a single data point, judgment must still be exercised in their use. The analyst may choose to use the generic data directly if the equipment description, process conditions, and failure mode of the data sources are similar to the equipment being studied. More likely, the analyst will have to adjust the data to account for differences in equipment design, process conditions, properties of the chemicals being processed, severity of duty, and/or quality of the facility maintenance regime.

This chapter has discussed some of the factors that may affect equipment reliability and necessitate data adjustment. At this time, little documented assistance is available to help develop these data adjustments. It may be necessary to get help from experts in some situations. Lastly, failure rates are often reported to several decimal places, a precision frequently unwarranted by the data. It is suggested that only the failure rate's first significant number and associated exponential power be used.

References

1. *Guide to the Collection and Representation of Electrical, Electronic, Sensing Component, and Mechanical Equipment Reliability Data for Nuclear Generating Stations.* IEEE Std. 500-1984, Institute of Electrical and Electronic Engineers, New York, 1984.
2. Drago, J. P., Borkowski, R. J., Pike, D. H., and Goldberg F. F. *The In-Plant Reliability Data Base for Nuclear Power Plant Components: Data Collection and Methodology Report.* NUREG/ CR-2641, ORNL/TM-9216, January 1985.
3. *Offshore Reliability Data Handbook,* OREDA-84. P.O. Box 370, N-1322, HOVIK, Norway, 1984. Distributed by Pennwell Publishing Company, Tulsa, OK.
4. *Reliability Data Book for Components in Swedish Nuclear Power Plants*, RKS 85-25, Nuclear Safety Board of Swedish Utilities, Swedish Nuclear Power Inspectorate, Stockholm

3
CCPS Taxonomy

This chapter reviews the structure, rationale, and method used for the development of the CCPS Taxonomy and explains how to use it. Key elements of the CCPS Taxonomy that are explained include equipment, service, and failure description. The CCPS Taxonomy is listed in Appendix A.

3.1 CCPS Taxonomy Structure

The hierarchical structure of the CCPS Taxonomy is divided into three major parts: equipment description, service description, and failure description. Figure 3.1 illustrates this organization.

3.1.1 Equipment Description

As indicated in Figure 3.1, there are several levels within the equipment description. Each data cell has a unique taxonomy number, which is determined by the number of levels and its position within each level. For the example highlighted by shaded areas in Figure 3.1, the taxonomy number 3.3.7.2.1.1 specifies the third element of the first level, the third element of the second level, the seventh element of the third level, etc.

The upper two or three levels of the equipment description broadly categorize the equipment by generic type. Table 3.1 summarizes these first few levels for the entire CCPS Taxonomy. The lower levels of the equipment description are based on factors that have a strong influence on reliability. For example, the reliability of piping is a function of its diameter. Therefore, pipe diameter, one of the equipment description levels under piping, is divided into two groups: 6" or smaller; and larger than 6 ". Factors that may define levels within the equipment description include function, drive type, fabrication technique, internals, materials of construction, and design principle. These lower levels are important because they further define the data cells, and, if chosen properly, can reduce the scatter of the failure rate data. The number and nature of the levels varies with each data cell.

The final element of the equipment description is the equipment boundary figure. A boundary figure is included with each data cell to define the components and limits of the equipment associated with that cell. For example, the data cell boundary figure (Data cell 3.3.7.2.1.1) in Figure 3.2 shows that the centrifugal pump, seal system, motor, motor control unit, lube oil system, coupling, and transmission are all components of the equipment in the data cell. The equipment boundary is inherently reflected in the taxonomy number.

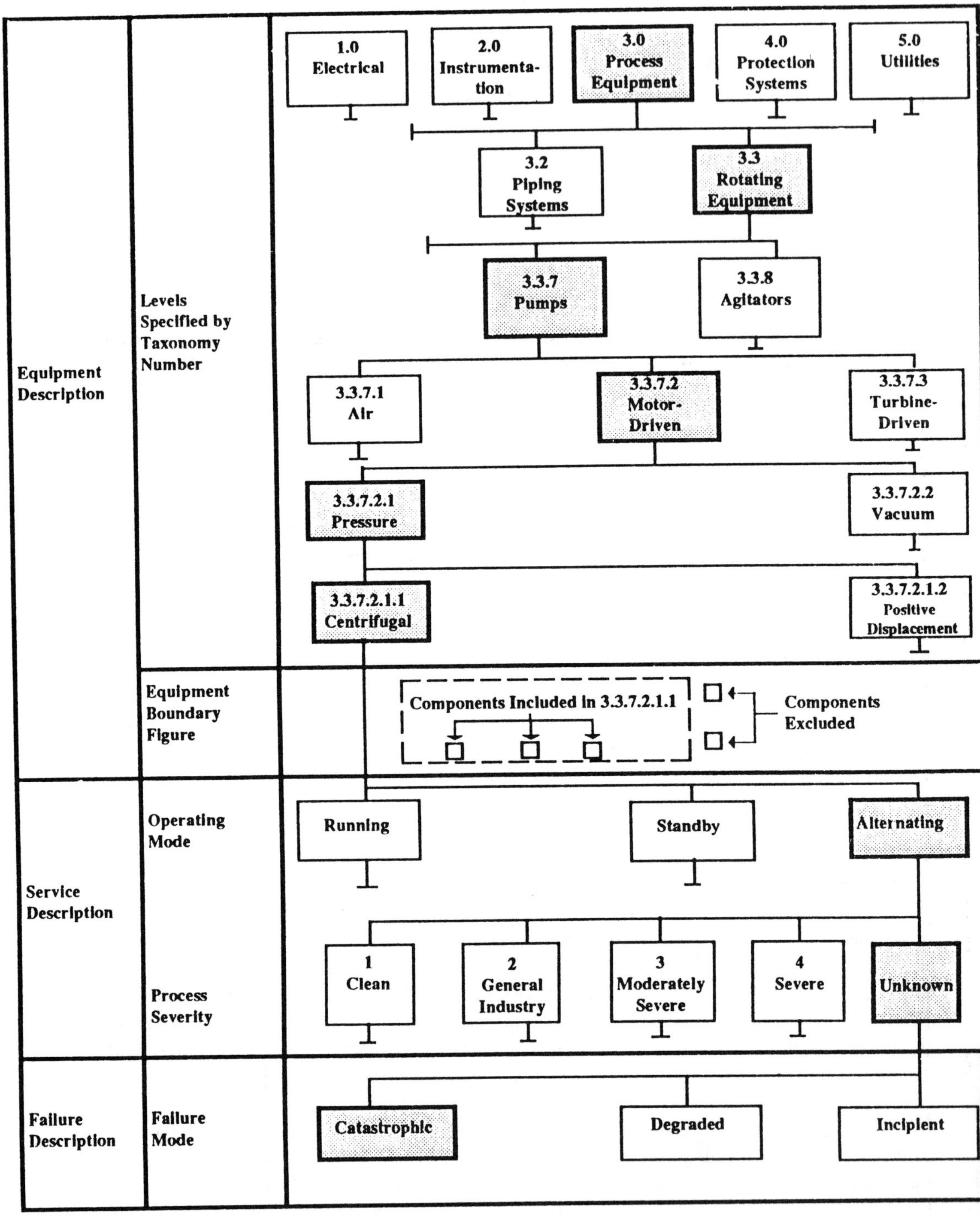

Figure 3.1. *Example of CCPS Taxonomy structure. The shaded areas illustrate the taxonomy number, 3.3.7.2.1.1, used as an example throughout this chapter.*

TABLE 3.1
Upper Levels of CCPS Taxonomy

- 1.0 Electrical equipment
  - 1.1 Motors
  - 1.2 Power conditioning and protection devices
  - 1.3 Power generation
- 2.0 Instrumentation
  - 2.1 Process wetted and field instrumentation
  - 2.2 Control room instrumentation
- 3.0 Process equipment
  - 3.1 Heat transfer devices
    - 3.1.1 Fired
    - 3.1.2 Non-fired
  - 3.2 Piping systems
    - 3.2.1 Metal
    - 3.2.2 Lined pipe
    - 3.2.3 Rigid plastic piping
    - 3.2.4 Tubing systems
    - 3.2.5 Hoses
  - 3.3 Rotating equipment
    - 3.3.1 Centrifuges
    - 3.3.2 Compressors
    - 3.3.3 Blowers
    - 3.3.4 Motor driven fans
    - 3.3.5 Extruders
    - 3.3.6 Mixers/blenders
    - 3.3.7 Pumps
    - 3.3.8 Rotary agitators
  - 3.4 Solids handling
    - 3.4.1 Baggers/packagers
    - 3.4.2 Conveyors
    - 3.4.3 Elevator
    - 3.4.4 Feeders
    - 3.4.5 Separators
    - 3.4.6 Size reducers
  - 3.5 Valves
    - 3.5.1 Check valves
    - 3.5.2 Manual valves
    - 3.5.3 Operated valves
  - 3.6 Vessels and accumulators
    - 3.6.1 Atmospheric
    - 3.6.2 Pressurized
    - 3.6.3 Vacuum
  - 3.7 Miscellaneous
    - 3.7.1 Electrolytic cells
    - 3.7.2 Seals/gaskets
- 4.0 Protection systems
  - 4.1 Corrosion
  - 4.2 Fire
  - 4.3 Pressure
- 5.0 Utilities
  - 5.1 Cooling water systems
  - 5.2 Flares
  - 5.3 Gas generators
  - 5.4 Heating systems
  - 5.5 Heating, ventilating and air conditioning (HVAC)
  - 5.6 Incinerators
  - 5.7 Refrigeration
  - 5.8 Steam systems

DATA ON SELECTED PROCESS SYSTEMS AND EQUIPMENT	
Taxonomy No. 3.3.7.2.1.1	**Equipment Description** ROTATING EQUIPMENT- PUMPS MOTOR DRIVEN-PRESSURE-CENTRIFUGAL
Operating Mode ALTERNATING	**Process Severity** UNKNOWN

Population	Samples	Aggregated time in service (10^6 hrs)		No. of Demands		
		Calendar time	Operating time			
Failure mode	Failures (per 10^6 hrs)			Failures (per 10^3 demands)		
	Lower	Mean	Upper	Lower	Mean	Upper
CATASTROPHIC						
a. Fails while Running	43.3	292.0	862.0			
b. Rupture						
c. Spurious Start						
d. Fails to Start on Demand				0.360	10.80	43.0
e. Fails to Stop on Demand						
DEGRADED						
a. Fails to Run at Rated Speed	15.8	920.0	3560.0			
b. External Leak						
INCIPIENT						
a. High Vibration						
b. Over-temperature						
c. Over-current						

Equipment Boundary

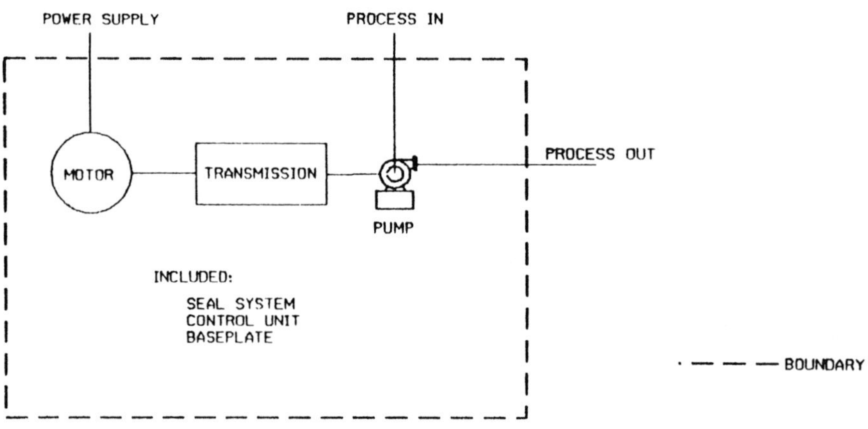

Data Reference No. (Table 5.1): 5, 8.1

Figure 3.2. *Example of CCPS generic data sheet.*

Many data cells in the CCPS Taxonomy use equipment boundaries found in available generic data sets in which equipment and service is similar to that in the CPI. The boundaries established for other data cells were generally combinations of normal equipment modules—such as pump, seals, coupling, motor and base plate, or refrigeration units—and functionally interdependent basic and auxiliary components, such as motor controllers. Boundaries may change as greater amounts of equipment reliability data become available.

3.1.2 Service Description

The service description is the second part of the taxonomy structure. It includes two levels—the operating mode and the process severity. The operating mode describes how the equipment is operated. The three modes used, running, standby, and alternating, are defined as follows:

- **Running:** Hardware that is usually operating (e.g., an unspared compressor that must operate to run the process).
- **Standby:** Hardware that is normally not running, but must be ready to run (e.g., an emergency diesel generator).
- **Alternating:** Hardware that alternates between standby and running (e.g., a pump and installed spare running a comparable amount of time).

The process severity characterizes the process medium, the material being handled by the equipment. There are four categories in the taxonomy, defined as follows:

1. **Clean:** Clean fluids (e.g., dry clean air, potable water, nitrogen).
2. **General industry:** Noncorrosive, nonabrasive, and nonplugging fluids (e.g., natural gas, ethanol, general service steam).
3. **Moderately severe:** Moderately corrosive, moderately abrasive or moderately plugging (e.g., dry chlorine, anhydrous ammonia, untreated sea water).
4. **Severe:** Severely corrosive, severely abrasive or severely plugging (e.g., wet hydrogen chloride, coal slurry, heavies, uninhibited monomer).

3.1.3 Failure Description

The failure description is the third part of the taxonomy structure and involves the modes, severities, and types of failures. These are based on models in the *In-Plant Reliability Data Base for Nuclear Power Plant Components: Data Collection and Methodology Report* (IPRDS)[1] and IEEE Std. 500-1984,[2] which are discussed in Chapter 2.

3.2 CCPS Taxonomy Development

The development of the CCPS Taxonomy was essential, and it is one of the most useful results of preparing this book. The first step in creating the taxonomy was to develop a list of equipment to be included. This was accomplished by reviewing the equipment lists of over 300 different chemical processes summarized by SRI International in their Process Economics series. These lists were used in conjunction with the CCPS Equipment Reliability Data Subcommittee's experience to establish a list of important CPI equipment for inclusion in the taxonomy.

The equipment was then categorized by generic type into the upper levels of the equipment description (see Table 3.1). Each type of equipment was further reviewed in order to establish the lower levels of the equipment description. Factors that were judged to be important distinguishers of reliability were used to define additional levels in the CCPS Taxonomy. Some obvious equipment characteristics were not used as equipment descriptors in the taxonomy because experience has shown they have little influence on reliability behavior. For example, valves were not characterized by type (globe, gate, plug, etc.). In this case, more important distinguishers of reliability are means of operation, such as manual or automatic.

Table 3.2 is a list of factors that were generally not used to establish hierarchical levels in the CCPS Taxonomy. In some cases, the effect of a factor (e.g., humidity) on equipment reliability was insufficient to warrant a new taxonomy level. In many others, criteria established by the subcommittee eliminated them as taxonomy levels. These criteria assumed proper equipment design and specification for the service (e.g., materials of construction, pressure rating, shop fabrication), good installation practices and maintenance programs, and no unusual operating stresses (operating at less than 100% of design).

Many of the data cells defined by the CCPS Taxonomy are not presented in Chapter 5 because no appropriate data were available. It is hoped that this book will promote the development of new data by the CPI to fill these empty cells. As new data are collected, modifications to the CCPS Taxonomy may be required to better reflect reliability influences. The new data should provide better answers to the following questions:

- What equipment should be included in the taxonomy?
- What are the most important distinguishers of reliability?
- What are the proper service descriptions and failure modes of the equipment?
- What are the proper equipment boundaries?

3.3 The CCPS Taxonomy and Its Use

The development of the CCPS Taxonomy provides an enormous number of data cells, each with its unique taxonomy number. To conserve space, the CCPS Taxonomy has been condensed for presentation in Appendix A. Figure 3.3 is an example page from Appendix A.

TABLE 3.2
Factors Not Assigned as Taxonomy Levels

1. External environment, for example, vibration, temperature, humidity.
2. Fabrication techniques, inspection procedures, installation practices (except as noted).
3. Materials of construction (except as noted).
4. Maintenance strategy.
5. Service stress, for example, heavy, medium or light duty.
6. Internal temperature or pressure (except as noted).
7. Design standards.
8. Process type or equipment manufacturer.

3.0 **PROCESS EQUIPMENT**
3.3 **Rotating Equipment**

APPENDIX A - CCPS TAXONOMY

EQUIPMENT DESCRIPTION			SERVICE DESCRIPTION		FAILURE DESCRIPTION
			OPERATING MODE	PROCESS SEVERITY	
3.3.7 **Pumps** 3.3.7.1 Air		.1 0 - 1000 HP .2 >1000 HP	Alternating Running Standby	1 - 4	1- Catastrophic a. Fails While Running b. Rupture c. Spurious Start d. Fails to Start on Demand e. Fails to Stop on Demand 2- Degraded a. Fails to Run at Rated Speed b. External Leak 3- Incipient a. High Vibration b. Over-temperature c. Over-current
3.3.7.2 Motor-Driven 3.3.7.2.1 Pressure	.1 Centrifugal .2 Positive Displacement .2.1 Gear .2.2 Piston				
3.3.7.2.2 Vacuum	.1 Centrifugal .2 Positive Displacement				
3.3.7.3 Turbine-Driven 3.3.7.3.1 Gas 3.3.7.3.2 Steam					
3.3.8 **Rotary Agitators** 3.3.8.1 Direct-Driven 3.3.8.2 Gear-Driven			Alternating Running	2 - 4	1- Catastrophic a. Fails While Running b. Seal External Rupture c. Seal Rupture - Influx d. Spurious Start/Command Fault e. Fails to Start on Demand f. Fails to Stop on Demand

Figure 3.3. *Example of condensed CCPS Taxonomy.*

TABLE 3.3
Expansion of the CCPS Taxonomy Example, 3.3.7.2.1.1
(see Figures 3.2 and 3.3)

3 Process equipment
3.3 Rotating equipment
3.3.7 Pumps
3.3.7.2 Motor driven
3.3.7.2.1 Pressure
3.3.7.2.1.1 Centrifugal

Equipment Description	Service Description: Operating Mode	Service Description: Process Severity	Failure Description
3.3.7.2.1.1(0-1000 HP)	Alternating	1	*
"	"	2	*
"	"	3	*
"	"	4	*
"	Running	1	*
"	"	2	*
"	"	3	*
"	"	4	*
"	Standby	1	*
"	"	2	*
"	"	3	*
"	"	4	*
3.3.7.2.1.1(>1000HP	Alternating	1	*
"	"	2	*
"	"	3	*
"	"	4	*
"	Running	1	*
"	"	2	*
"	"	3	*
"	"	4	*
"	Standby	1	*
"	"	2	*
"	"	3	*
"	"	4	*

*Failure modes:

1. Catastrophic
   a. Fails while running
   b. Influx of contaminants (backflow)
   c. Spurious start
   d. Fails to start on demand
   e. Fails to stop on demand
2. Degraded
   a. Fails to run at rated speed
   b. External leak
3. Incipient
   a. High vibration
   b. Over-temperature
   c. Over-current

As shown in this figure, the format is divided into three main columns labeled "Equipment Description," "Service Description," and "Failure Description." The Equipment Description column may be further divided to show the necessary equipment description levels that make up the taxonomy number. Each column represents one additional hierarchical level and number in the CCPS Taxonomy. Similarly, the Service and Failure Descriptions are divided as needed to fully establish the data cells. An entry or group of entries in a column apply all the way down the column until an additional entry or a horizontal line is reached.

To illustrate the effect of condensing the taxonomy for Appendix A, Table 3.3 shows the expansion of taxonomy number 3.3.7.2.1.1, one part of the taxonomy in Figure 3.3. There are 24 potential data cells in Table 3.3. The first data cell listed has a taxonomy number of "3.3.7.2.1.1," an "Alternating" Operating Mode and a Process Severity of "1."

When trying to locate data on a specific type of equipment in the CCPS Generic Failure Rate Data Base, the reader will need its taxonomy number. Table 3.1 is useful for determining the section of the taxonomy in Appendix A where it can be found. Alternatively, the taxonomy number can be located directly by referring to Appendix B, the Equipment Index with associated taxonomy numbers. To help the reader, Table 5.2 lists by taxonomy number those cells which contain data and are in the CCPS Generic Failure Rate Data Base.

References

1. Drago, J. P., Borkowski, R. J., Pike, D. H., and Goldberg, F.F. *The In-Plant Reliability Data Base for Nuclear Power Plant Components: Data Collection and Methodology Report.* NUREG/CR-2641, ORNL/TM-9216, January 1985.
2. *Guide to the Collection and Representation of Electronic, Sensing Component, and Mechanical Equipment Reliability Data for Nuclear Generating Stations.* IEEE Standard 500-1984, Institute of Electrical and Electronics Engineers, New York, 1984.
3. *EuReDatA Project Report No. 1, Reference Classification Concerning Components Reliability.* T. Luisi, Commission of the European Communities, Joint Research Centre, Ispra Establishment, S.A./1.05.01.83.02.
4. *EuReDatA Project Report No. 3, Guide to Reliability Data Collection and Management.* B. Stevens, Commission of the European Communities, Joint Research Centre, Ispra Establishment, S.P./1.05.E3.86.20
5. *Offshore Reliability Data Handbook,* OREDA-84. P.O. Box 370, N-1322, HOVIK, Norway, 1984. distributed by Pennwell Publishing Company, Tulsa, OK.

4
Data Bases, Sources, and Studies

This chapter provides summaries of selected data resources available to the CPQRA practitioner. These resources are summarized in a consistent format that allows them to be easily reviewed and compared. Those resources which are available to CCPS and contain equipment failure rate data of sufficient quality are used for the data tables in Section 5.5.

Section 4.1 describes how data resources were chosen for inclusion in this book. Section 4.2 describes the format used to present the information on the data resources, and Sections 4.3 through 4.8 present these data resources. Each resource section is preceded by an index of the resources presented in the section.

4.1 Data Resource Selection

The selection of data resources was a three-step process:

1. Titles of potential resources were obtained by conducting a literature search and an industry survey. Simultaneous literature searches were conducted by CCPS and SAIC. CCPS concentrated on obtaining CPI data resources while SAIC used a literature search conducted for the nuclear power reliability community. These literature searches used in-house company, engineering, and public libraries and recommendations from members of the user community. At the same time, a questionnaire was sent to professionals who conduct CPQRAs. The survey requested information on the data resources used by the companies and whether they had plant-specific data that could be used by CCPS. Members of the CCPS Equipment Reliability Data Subcommittee were also asked to compile lists of data resources with which they were familiar and which they had used for reliability or risk analyses. As a result, an extensive but not necessarily complete list of data resource titles was assembled. Any resources uncovered after the publisher's cutoff date and not reviewed have been included in Appendix D.

The effort to collect CPI and general reliability data resources is considered by CCPS to be an ongoing project. Users of this book are encouraged to assist in this process by recommending additional resources to CCPS that can be used for subsequent editions of this book.

2. Subcommittee members selected those resources for further study which had titles suggesting that the resource might contain equipment failure rate data. Copies of these resources were then obtained and read.

3. Summaries of the data resources considered useful were prepared. "Useful data" was defined as information that was publicly available, scientifically collected, had statistical merit, and could be used for CPQRAs. A list of rejected resources was retained to identify references for supplemental reading and to avoid review duplication when the anticipated second edition of this book is developed. In total, 72 resources were accepted, and over 200 references were rejected.

The selected data resources were sorted into the six categories, each presented in a section of this chapter. Resources are numbered consecutively within each category. The sections and categories are:

Section 4.3 Process Equipment Data Bases
Section 4.4 Process Equipment Data Sources
Section 4.5 Chemical Process Quantitative Risk Analyses (CPQRA)
Section 4.6 Nonprocess Equipment Data Bases
Section 4.7 Nonprocess Equipment Data Sources
Section 4.8 Nuclear Probabilistic Risk Assessments (PRA)

The terms used in these categories are defined below:

Process: Refers to the CPI.

Nonprocess: Refers to industries that do not comprise a part of the CPI as their primary function, but which use comparable or equivalent complex equipment systems to perform their function, such as nuclear power plants, fossil fuel plants, and offshore oil rigs.

Data base: A repository for equipment reliability information categorized to facilitate data retrieval; or tabular lists of multiple data vectors, with little text except that needed to explain the data presentation format.

Data source: Descriptive text in a given subject area whose primary purpose is to discuss a reliability or risk topic but which also contains some useful reliability data.

Risk study (CPQRA or PRA): Specific study performed on a particular facility to determine the areas of weakness and strength in equipment and plant performance reliability; may include consequence analysis and usually implies some judgment of the risk.

The contents of these data resource summaries have been verified by calling the contacts and confirming ordering addresses. Information is current as of August 1988.

4.2 Data Resource Presentation

This section describes the format used to characterize the information in the data resources presented in Sections 4.3 through 4.8. The following data elements are used to present information in the resources summaries:

Title: The name of the data resource as shown on the data source, as described in literature explaining data base services, or as conventionally cited when referencing a CPQRA or PRA.

Sponsor/author: The organization or individual(s) responsible for performing or funding the collection and analysis of the information in the resource.

Number: An identifying number based on the resource category and its sequential placement within the category; in the range of 4.3-x to 4.8-x.

Industry: The industry from which the data originated, such as chemical process, power, nuclear, and offshore oil.

Time frame: The years (and months, where available) of calendar period covered by the information in the resource.

Type: The nature of the resource: data base, report, or paper.

Frequency of update: The regularity of incorporation of new information into the resource or the rate of issuance of new editions of the resource.

Number and type of record: The number of data points or tables of data presented in the resource or the number of events the data set reflects; where available, the form in which the data are presented, such as failure rates or availability data, confidence intervals or error factors; the "raw" data source used, such as surveys, plant records, tests, or judgment.

Data boundary: The equipment types for which data are presented in the resource (such as pipelines, valves or instruments); plant type (coal gasification facility, nuclear or ammonia plant, for example).

Data access: The contact for technical information, the document ordering address and phone number, the report number to use when ordering, the cost where available, and the data or report accessibility, if relevant.

Description: A summary of the content of the resource, some history of its development and the developers, and further information on the other fields in the presentation format.

A sample page is included from many of the data bases.

To help the reader select the appropriate data resource, an index precedes Sections 4.3 through 4.8. The index provides the source number within the section and the following set of data elements for each source: title, industry, number and type of records, and data boundary. Appendix C contains additional information about the data elements presented in each data resource. It can also be used to help identify the resources which may provide data for a CPQRA. A discussion of the Appendix C Matrix and an explanation of data elements indexed is presented. After examining Appendix C and the pattern of data elements contained in the data resources, it is evident that equipment reliability data have been published in a variety of formats, often without any apparent effort to conform to a recognized standard for data specification. The CCPS Taxonomy and the raw data collection requirements in Chapter 6 present the basis for reliability data specification in future literature.

4.3 INDEX OF PROCESS EQUIPMENT DATA BASES

NO.	TITLE	INDUSTRY	NO. & TYPE OF RECORDS	DATA BOUNDARY	PAGE
4.3-1	Imperial Chemical Industries Reliability Data Book	Chemical Process	About 1,000 miscellaneous failure rates, event rates, and probabilities. There is some treatment of human error probability	Pumps, valves, pipes, bursting discs, human errors, and other miscellaneous items	31.
4.3-2	Development of an Improved Liquified Natural Gas Plant Failure Rate Data Base	Chemical Process	Mean Time Between Failures (MTBFs) based on 1,116 failures	21 system/component categories such as cryogenic valves, heat exchangers, and fire protection	32.
4.3-3	COMPI: Data Bank for Component Failure Data	Varied	Unknown number of records coded by component, failure mode, and data source	Mechanical, electrical, and electronic component failure data	34.
4.3-4	Fluor Daniel Inc. Data Base	Chemical Process, Power, Petrochemical, Telecommunications, Nuclear Fuel Cycle	Failure and restore data for specific and generic application	Varies extensively for particular industry sector some of which is specialized or unique	36.
4.3-5	Computerized Library of Equipment Failures (CLEF)	Chemical Process, Petroleum, Nuclear	1000+ failure rate entries from various published sources	The Database contains failure rate data for most major equipment items that are found throughout the process industries	38.
4.3-6	HARIS-Hazards and Reliability Information System-Reliability Data Base	Chemical Process, Petroleum, Natural Gas and Nuclear	Failure rate data from public domain sources and derived from field failure studies. Over 1,500 failure rates.	The data base contains failure rate data plus some failure mode information for process equipment - pumps, compressors, gas turbines, valves, vessels, heat exchangers etc.	40.

PROCESS EQUIPMENT DATA BASES

TITLE:
Imperial Chemical Industries Reliability Data Book

SPONSOR/AUTHOR: Imperial Chemical Industries	**NO.:** 4.3-1
INDUSTRY: Chemical Process	**TIME FRAME:** Varies
TYPE: Data Base and Report	**FREQUENCY OF UPDATE:** Infrequent

NUMBER AND TYPE OF RECORDS: About 1,000 miscellaneous failure rates, event rates, and probabilities. There is some treatment of human error probability

DATA BOUNDARY: Pumps, valves, pipes, bursting discs, human errors, and other miscellaneous items

DATA ACCESS:

Contact: Process Safety Section
ICI Engineering Department, P.O. Box 7, Northwich Cheshire CW8 4DJ
Phone: England 0606-704995, 0606-704712 (secretary)
Report accessibility: Available as part of 2-week risk assessment and reliability engineering training course for the process industries. Details from the above address or Health and Environmental Affairs, ICI America Inc., Wilmington, DE 19897. Phone: (302) 575-4501

DESCRIPTION:

The book contains, in alphabetical order, failure rates, event rates and probabilities, and descriptive information which has been collected since 1970 in the course of doing risk and reliability assessments. Twenty appendices contain results of surveys on bursting discs, pipes, valves, relief valves, pump failures and information on human error, international fire losses, and blast effects.

PROCESS EQUIPMENT DATA BASES

TITLE:
Development of an Improved Liquified Natural Gas Plant Failure Rate Data Base

SPONSOR/AUTHOR:
Gas Research Institute

NO.:
4.3-2

INDUSTRY:
Chemical Process

TIME FRAME:
March 1980 to June 1981

TYPE:
Data Base and Report

FREQUENCY OF UPDATE:
Intended every five years

NUMBER AND TYPE OF RECORDS: Mean Time Between Failures (MTBFs) based on 1,116 failures

DATA BOUNDARY: 21 system/component categories such as cryogenic valves, heat exchangers, and fire protection

DATA ACCESS:

Contact: Steve Wiersma
Gas Research Institute
8600 West Bryn Mawr Ave., Chicago, IL 60631
Phone: (312) 399-8100
Report order address: NTIS, Springfield, VA 22161
Phone: (703) 487-4650
NTIS Report No: PB 82-153503

DESCRIPTION:

Failure rate questionnaires were sent to 35 companies which operate LNG base loading or satellite facilities. These operators had previously expressed an interest in participating in the study. Twenty-five companies returned questionnaires which covered failures at 27 separate LNG facilities. Approximately 1,626,000 hours of plant operating time were represented by the returned questionnaires.

The results of the study are presented for these equipment groups: gas pretreatment systems, heat exchangers, vaporizers, cryogenic storage tanks, compressor systems, cryogenic pumps, cryogenic valves, cryogenic piping, piping insulation, equipment insulation, process control systems, human errors, spills and leaks, truck loading and unloading facilities, fire protection systems, hazard detection system.

Major and minor type failures are treated. Minor failures are defined as those which cause (or would have caused) an unscheduled shutdown of equipment for a period of less than 24 hours. A major failure is defined as any failure which results in an unscheduled shutdown for a period of greater than 24 hours. Safety-related failures were defined as failures which resulted either in a fire, injury, loss of life, or a large leak of liquid or gas. To qualify as a safety-related failure, the liquid or gas release had to be large enough to have the potential to injure or have injured plant personnel, or have been severe enough to propagate beyond the immediate area.

Data Resource 4.3-2

Development of an Improved Liquified Natural Gas Plant Failure Rate Data Base

Example Data Sheet

TABLE 1

SUMMARY OF MAJOR FAILURES

Plant Area	Operating Hours	Major Failures	MTBF (hours)
Gas Pretreatment	675,000	25	27,000
Heat Exchangers	2,837,000	16	177,000
Vaporizers	188,000	26	7,200
Cryogenic Storage Tanks	1,809,000	2	904,500
Cryogenic Storage Systems	1,809,000	4	452,000
Compressor Systems	2,256,000	116	19,000
Cryogenic Pumps	366,000	86	4,000
Cryogenic Valves	6,278,000	4	1,569,000
Cryogenic Piping	1,164,000,000*	2	582,000,000*
Piping Insulation	SD	SD	SD
Equipment Insulation	SD	SD	SD
Process Control Systems	1,505,000	9	167,000
Human Errors	4,779,000#	19	252,000#
Spills and Leaks	1,626,000	11	148,000
Truck Loading and Unloading	1,156,000	0	>1,156,000
Fire Protection Systems	1,450,000**	24[n]	60,000**
fire water systems	1,450,000**	14	104,000**
dry chemical systems	1,423,000**	2	712,000**
gas systems	364,000**	2	182,000**
foam systems	88,000**	0	>88,000**
Hazard Detection Systems	16,703,000	76[n]	220,000
gas detectors	16,703,000	44	380,000 (SD)
low temp. det.	2,631,000	2	1,315,000
flame det.	10,570,000	12	881,000
high temp. det.	8,418,000	0	>8,418,000

*ft-hours #operator-hours **in service hours [n]normalized

SD= see discussion

From Gas Research Institute GRI-80/0093

PROCESS EQUIPMENT DATA BASES

TITLE:
COMPI: Data Bank for Component Failure Data

SPONSOR/AUTHOR:
TNO

NO.:
4.3-3

INDUSTRY:
Varied

TIME FRAME:
1978 to Present

TYPE:
Data Base and Report

FREQUENCY OF UPDATE:
Continuous

NUMBER AND TYPE OF RECORDS: Unknown number of records coded by component, failure mode, and data source

DATA BOUNDARY: Mechanical, electrical, and electronic component failure data

DATA ACCESS:

Contact: TNO, Dept. of Industrial Safety, Ir. J. Van der Horst
P.O. Box 342, 7300 AH Apeldoorn, The Netherlands
Phone: (0) 55 77 33 44
Telex: 36395 TNOAP NL
Cost: Fixed price of 150 DFL, plus charge per hour for data search/analysis
Report accessibility: Letter requests specifying data searches needed

DESCRIPTION:

The Department of Industrial Safety has been collecting and recording component failure data since 1978. For this purpose use is made of the following sources of information: accessible (international) data-banks, literature and data from TNO research projects. The Department of Industrial Safety has failure data-bases relating to: mechanical components, electronic components, and electric components.

As failure data relating to mechanical components differ widely from source to source, TNO has set up a documentation system in which all relevant information is stored in one, uniform automated code called COMPI, which uses a component description code for the following information: system of construction, operation and function. The automated data-base supplies the following standard output: component code; failure behaviour, cause of failure; failure rate, i.e. the quotient of the number of failures and the number of service hours/cycles; and source of information. On request further particulars can be supplied like: (a) failure rate: number of failures, number of service hours, calculated distribution. (b) the conditions and situations in which the component has been used (e.g. the medium: LNG, ammonia, and petrol).

The failure data relating to electronic and electric components are available in the form of handbooks. Failure rates are derived with the aid of calculation models based on statistical relations for which the incorporation of a (large) number of parameters is required. The following minimum of information is needed: type of component, manufacturer and environmental factors.

Data Resource 4.3-3
COMP 1: Data Bank for Component Failure Data
Example Data Sheet

```
Componenttype : turbines,                    Componentcode :  90140,
Par1 : type : 35,                            Par2 : - :  0,
Par3 : aantal stappen :  0,                  Par4 : rotatiesnelheid :  0,
Par5 : druk :  0,                            Par6 : vermogen :  0,
F(ailure)mode : -,                           Informationtype : - ,
Failure rate : 2,5*10-2F/HR M,               Literatuurreferentie : E1,
Divinf : - : -,                                          Refnr :      1
```

E1
Earles D.R.

Reliability Application and Analysis Guide MI-60-54

The Martin Company, july 1961

```
Componenttype : turbines,                    Componentcode :  90140,
Par1 : type : 35,                            Par2 : - :  0,
Par3 : aantal stappen :  0,                  Par4 : rotatiesnelheid :  0,
Par5 : druk :  0,                            Par6 : vermogen :  0,
F(ailure)mode : -,                           Informationtype : - ,
Failure rate :   .4        f/yr    M,        Literatuurreferentie : M3,
Divinf : - : -,                                          Refnr :      3
```

```
Componenttype : turbines,                    Componentcode :  90140,
Par1 : type : 80,                            Par2 : - :  0,
Par3 : aantal stappen :  0,                  Par4 : rotatiesnelheid :  0,
Par5 : druk :  0,                            Par6 : vermogen :  0,
F(ailure)mode : -,                           Informationtype : - ,
Failure rate :   .6        f/yr    M,        Literatuurreferentie : M3,
Divinf : - : -,                                          Refnr :      4
```

M3
Moss T.R.

Plant Availability Assessment

NCSR R13, December 1978

Reprinted with permission of TNO, Department of Industrial Safety

PROCESS EQUIPMENT DATA BASES

TITLE:

Fluor Daniel Inc. Data Base

SPONSOR/AUTHOR:

Various

NO.:

4.3-4

INDUSTRY: Chemical Process, Power, Petrochemical, Telecommunications, Nuclear Fuel Cycle

TIME FRAME:

1975 to Present

TYPE:

Data Base

FREQUENCY OF UPDATE:

With new data availability

NUMBER AND TYPE OF RECORDS: Failure and restore data for specific and generic application.

DATA BOUNDARY: Varies extensively for particular industry sector some of which is specialized or unique.

DATA ACCESS:

Contact: RAM Engineering (B4H)
Fluor Daniel Inc. 3333 Michelson Drive, Irvine, CA 92730
Phone: (714) 975-5854
Report ordering address: Same as above
Report cost: Varies with consulting time required to provide relevant data.
Report accessibility: Through data consulting agreement.

DESCRIPTION:

Fluor Daniel Inc. has developed data that it applies to a wide spectrum of industry applications. This data is updated and tailored for specific use requirements necessary to perform risk and availability assessments. Maintenance and operational factors are assessed and incorporated for specific uses.

Data Resource 4.3-4
Fluor Daniel Data Base
Example Data Sheet

TABLE 1-8 DATA

CMO Unit Component Data

Component	MTTF	MTTR
Feed Pumps	33000 hrs	27 hrs
Feed Preheater	100000	56
Main Feed Vaporizer	33000	56
Methanol Drum	91000	37
Dehydration Reactor	26300	106
Feed / Product Exchanger	58200	86
Product Condensers	100000	52
Three Phase Separator	91000	37
Aqueous Phase Pumps	33000	27
Gasoline Phase Pumps	33000	27
Olefin Compressor	11400	39
Interstage Separators	91000	37
Intercoolers	100000	56
Aftercooler	100000	56
Stabilizer	17500	47
Condenser	100000	54
Reflux Drum	91000	39
Reflux Pumps	33000	27
Overhead Gasoline Pumps	33000	27
Gasoline Trim Cooler	100000	56
Feed / Bottoms Exchanger	33000	58
Reboiler	33000	58
Gasoline Trim Cooler	100000	58

TABLE 1-9 DATA

Separation Unit Component Data

Component	MTTF	MTTR
Depropanizer	17500 hrs	47 hrs
Overhead Condenser (water)	100000	58
Overhead Condenser (refrig)	33000	58
Reflux Drum	91000	39
Reflux Pumps	33000	27
Reboiler	100000	58

PROCESS EQUIPMENT DATA BASES

TITLE:
Computerized Library of Equipment Failures (CLEF)

SPONSOR/AUTHOR:
Technica, and Various other sources

NO.:
4.3-5

INDUSTRY: Chemical Process Petroleum, Nuclear

TIME FRAME:
1980 to Present

TYPE:
DataBase

FREQUENCY OF UPDATE:
Continuous

NUMBER AND TYPE OF RECORDS: 1000+ Failure rate entries from various published sources

DATA BOUNDARY: The Database contains failure rate data for most major equipment items that are found throughout the process Industries.

DATA ACCESS:

Contact: Matthew J. Zerafa, Database Manager
Technica Inc.
Phone: (614) 848-4000
Report ordering address: 355 Campus View Blvd.,
Columbus, Ohio 43235
Report cost: A fee gives clients access to initial CLEF data set
Report accessibility: Available in computer form to clients

DESCRIPTION:

Technica has compiled computerized failure rate data from the public domain that can developed into a database. Each database can be customized by adding client plant-specific data and updated easily in its electronic form. CLEF is also software compatible with the IRRAS fault tree package put out by EG&G. Failure rate libraries can be generated and imported from CLEF to the IRRAS program.

Data Resource 4.3-5

Computerized Library of Equipment Failures

Example Data Sheet

COMPONENT : CIRCUIT BREAKERS * *

FAILURE MODE : ALL MODES

FAILURE RATES LOW	MEDIAN	UPPER	AVERAGE REPAIR	TEST INTERVAL	REFERENCE SOURCE
20.0e-003	35.0e-003	70.0e-003	0.0e+000	0.0e+000	OREDA
0.0e+000	20.0e-003	0.0e+000	0.0e+000	0.0e+000	CONF A
17.0e-005	20.0e-004	49.0e-001	0.0e+000	0.0e+000	IEEE-500

FAILURE MODE : FAIL TO CLOSE

FAILURE RATES LOW	MEDIAN	UPPER	AVERAGE REPAIR	TEST INTERVAL	REFERENCE SOURCE
0.0e+000	10.0e-003	0.0e+000	0.0e+000	0.0e+000	LEES
0.0e+000	26.0e-005	0.0e+000	0.0e+000	0.0e+000	IEEE-500
0.0e+000	17.0e-004	0.0e+000	38.0e-005	0.0e+000	CONF A

FAILURE MODE : F TO TRANSFER (/D)

FAILURE RATES LOW	MEDIAN	UPPER	AVERAGE REPAIR	TEST INTERVAL	REFERENCE SOURCE
30.0e-005	10.0e-004	30.0e-004	0.0e+000	0.0e+000	WASH-1400

FAILURE MODE : FAIL TO TRIP

FAILURE RATES LOW	MEDIAN	UPPER	AVERAGE REPAIR	TEST INTERVAL	REFERENCE SOURCE
0.0e+000	50.0e-004	0.0e+000	0.0e+000	0.0e+000	CONF A
0.0e+000	17.0e-005	0.0e+000	0.0e+000	0.0e+000	IEEE-500

FAILURE MODE : SPURIOUS

FAILURE RATES LOW	MEDIAN	UPPER	AVERAGE REPAIR	TEST INTERVAL	REFERENCE SOURCE
0.0e+000	10.0e-004	0.0e+000	0.0e+000	0.0e+000	CONF A
0.0e+000	86.0e-006	0.0e+000	0.0e+000	0.0e+000	IEEE-500

FAILURE MODE : FAIL TO CLEAR

FAILURE RATES LOW	MEDIAN	UPPER	AVERAGE REPAIR	TEST INTERVAL	REFERENCE SOURCE
0.0e+000	30.0e-004	0.0e+000	0.0e+000	0.0e+000	CONF D

PROCESS EQUIPMENT DATA BASES

TITLE: HARIS - Hazards and Reliability Information System - Reliability Data Base

SPONSOR/AUTHOR: Various Sources

NO.: 4.3-6

INDUSTRY: Chemical Process, Petroleum, Natural Gas and Nuclear

TIME FRAME: 1979 to date

TYPE: Data Base and Report

FREQUENCY OF UPDATE: Continuous

NUMBER AND TYPE OF RECORDS: Failure rate data from public domain sources and derived from field failure studies. Over 1,500 failure rates.

DATA BOUNDARY: The data base contains failure rate data plus some failure mode information for process equipment - pumps, compressors, gas turbines, valves, vessels, heat exchangers etc.

DATA ACCESS:

Contact: HARIS Manager (Mr. P. Stead)
RM Consultants Ltd, Suite 7, Hitching Court
Abingdon, Oxon, OX14 1DY, ENGLAND
Phone: 0235 - 555755
Report cost: Based on time and type of data required
Report accessibility: Public domain source documents available - also abstracts

DESCRIPTION:

The RMC HARIS (Hazards and Reliability Information System) programs provide organizations with a data bank of reliability, maintainability, accident, and source-abstract data. The programs permit the input of information in a standard data sheet format. Search capability is built into the programs for retrieval of these data sheets against specific search profiles. HARIS presently contains over 4400 data sheets.

4.4 INDEX OF PROCESS EQUIPMENT DATA SOURCES

NO.	TITLE	INDUSTRY	NO. & TYPE OF RECORDS	DATA BOUNDARY	PAGE
4.4-1	Pressure Vessel Reliability	Chemical Process and Power	Disruptive failure probability using US and Foreign data	Power boilers and unfired pressure vessels per Sections I and VIII of ASME Code	42.
4.4-2	Safety of Interstate Natural Gas Pipelines	Power	87 Casualities/Approx. 3000 pipe failures	U.S. interstate Natural Gas large diameter, high pressure transmission pipelines	43.
4.4-3	Some Data on the Reliability of Pressure Equipment in the Chemical Plant Environment	Chemical Process	Data derived from 1.4×10^4 vessel-year	Process pressure vessels, pressure storage vessels, and heat exchangers	44.
4.4-4	Some Data on the Reliability of Instruments in the Chemical Plant Environment	Chemical Process	40 instrument failure rates	Maintenance records on 9,500 instruments	45.
4.4-5	Failure and Maintenance Data Analysis at a Petrochemical Plant	Chemical process	Mean-time-between-maintenance action for five classes of equipment	Ethylene plant pumps; ethylbenzene-styrene monomer plant gas compressors, screw conveyors, pumps, and other non-moving equipment	46.
4.4-6	Causes of Ammonia Plant Shutdowns: Survey V	Chemical Process	Records of 5884 shutdowns over 98 years operating time	Ammonia Plant	47.
4.4-7	Reliability Data Collection and Use in Risk and Availability Assessment	Varied	72 papers, several of which contain some data	Wide variety of systems and components	48.
4.4-8	Pipeline Reliability: An Investigation of Pipeline Failure Characteristics and Analysis of Pipeline Failure Rates	Petroleum and Natural Gas	Data on frequency and cause of pipeline failures	Data is specific to submarine and cross-country oil and natural gas pipelines	49.
4.4-9	Fault Tree Analysis Report for Coal-Gasification Process-Development Unit	Coal Gasification	Failure rates (per year basis) for over 400 events from fault trees; Unavailability data	Coal-gasification Process Development unit systems and components	50.
4.4-10	Data Base Development and Equipment Reliability for Phase 1 of the Probabilistic Risk Analysis DPST-87-642	Chemical Process, Nuclear Fuel Cycle	About 250 component failure rates and 95% upper bounds	Pumps, valves, pipe, motors, diesels, heat exchangers, relays, fans for systems.	51.
4.4-11	Reliability Analysis of Pumps for Uranium Solutions	Chemical Process, Nuclear	Operating life repair frequency and cost data	Dual-diaphragm pumps in uranium solution service	52.
4.4-12	Emergency Generators: A Reliability Study Based on an Analysis of Failures	Chemical Process	138 Emergency Generator Failures	Diesel engine driven, emergency generators	53.
4.4-13	Reliability of a Solids-Fluid Handling Process Plant	Ore Processing	Failure modes and failure rates covering 829 failures/repairs	Solids handling equipment for one plant: rotary kiln, leaching tank, screwfeeder, and associated items	54.
4.4-14	Reliability Assessment of Safety/Relief Valves	Chemical Process	866 records	Safety and Relief Valves	55.

PROCESS EQUIPMENT DATA SOURCES

TITLE:
Pressure Vessel Reliability

SPONSOR/AUTHOR:
S.H. Bush

NO.:
4.4-1

INDUSTRY:
Chemical Process & Power

TIME FRAME:
Varies

TYPE:
Report

FREQUENCY OF UPDATE:
None

NUMBER AND TYPE OF RECORDS: Disruptive failure probabilities using US and Foreign data.

DATA BOUNDARY: Power boilers and unfired pressure vessels per Sections I and VIII of ASME Code.

DATA ACCESS:

Contact: American Society of Mechanical Engineers (ASME)
345 E. 47th St., New York, NY 10017
Phone: (212) 705-7794

Order from: ASME Order Dept. Phone: (201) 882-1167

DESCRIPTION:

The purpose of this study was to examine limitations in US and foreign chemical process and power industry pressure vessel data to determine if it can be encoded. In Germany the Institut fur Reaktorsicherheit der Technischen Uberwachungs-Vereine, (TUV), requires data collection on pressure vessels. In the US, data from the American Boiler Manufacturers Association (ABMA) concerning the ASME Boiler and Pressure Vessel Code Section I Boilers is available but is not inclusive enough for data encoding. The US appears to have a lower incidence of operational failures than the foreign sources report but study of the variors data does not indicate why. Relevant comparative data exists on nuclear reactor pressure vessels.

PROCESS EQUIPMENT DATA SOURCES

TITLE:

Safety of Interstate Natural Gas Pipelines

SPONSOR/AUTHOR:
Federal Power Commission

NO.:
4.4-2

INDUSTRY:
Power

TIME FRAME:
1950 to 1965

TYPE:
Report

FREQUENCY OF UPDATE:
None

NUMBER AND TYPE OF RECORDS: 87 Casualities/Approx. 3000 pipe failures.

DATA BOUNDARY: U.S. interstate Natural Gas large diameter, high pressure transmission pipelines.

DATA ACCESS:

Contact: Available only through public libraries. The report is not retained by Government Printing Office

Report accessibility: Public Information; Recorded in the 2nd Session of the 89th Congress as a Report for the U.S. Senate Committee on Conmmerce, April 19, 1966.

DESCRIPTION:

Military/government type publication. It lists accidents with fatality scenarios that occurred during operation and maintenance of U.S. interstate gas pipelines from 1950 to 1965. Also listed are individual pipework failures during that time, about 3000 entries, that have been compiled from various sources by the study committee.

PROCESS EQUIPMENT DATA SOURCES

TITLE: Some Data on the Reliability of Pressure Equipment in the Chemical Plant Environment

SPONSOR/AUTHOR: D.C. Arulanantham & F.P. Lees

NO.: 4.4-3

INDUSTRY: Chemical Process

TIME FRAME: 1950's to 1970's

TYPE: Journal Article

FREQUENCY OF UPDATE: None

NUMBER AND TYPE OF RECORDS: Data derived from $1.4x10^4$ vessel-years.

DATA BOUNDARY: Process pressure vessels, pressure storage vessels, and heat exchangers.

DATA ACCESS:

Contact: INT. J. Pressure Vessel & Piping
Elsevier Applied Science Publishers LTD, Great Britain, 1981

Report accessibility: Available Through Library Sources.

DESCRIPTION:

This is a survey of pressure equipment failure rate data including pressure vessels and heat exchangers. Overall failure rates given as $4x10^{-3}$ f/yr with an upper bound of $6.3x10^{-3}$ f/yr at a 99% confidence level. No disruptive failures were recorded but an upper bound of $2.8x10^{-4}$ f/yr at 99% confidence level is reported. Other items covered in the survey include non-pressure storage vessels and fired heaters. The data are also analysed to determine the effect of operating conditions such as high and low temperature and corrosive environments.

PROCESS EQUIPMENT DATA SOURCES

TITLE:
Some Data on the Reliability of Instruments in the Chemical Plant Environment

SPONSOR/AUTHOR:
S. N. Anyakora, G. F. M. Engel and F. P. Lees

NO.:
4.4-4

INDUSTRY:
Chemical Process

TIME FRAME:
1968 to 1970

TYPE:
Journal Article

FREQUENCY OF UPDATE:
None

NUMBER AND TYPE OF RECORDS: 40 instrument failure rates.

DATA BOUNDARY: Maintenance records on 9,500 instruments.

DATA ACCESS:

Contact: The Chemical Engineer, November 1971, pgs. 396-402.

Report accessibility: Libraries

DESCRIPTION:

Data have been obtained from maintenance records on the reliability of some 9,500 instruments in three chemical works with total operating time of about 4,500 instrument years. From these, failure rates have been estimated for about 40 instrument types.

Some attempt has also been made to assess the impact of location and contact with process materials on the failure rates.

PROCESS EQUIPMENT DATA SOURCES

TITLE:
Failure and Maintenance Data Analysis at a Petrochemical Plant

SPONSOR/AUTHOR:
D. J. Sherwin

NO.:
4.4-5

INDUSTRY:
Chemical Process

TIME FRAME:
Before 1982; 10 months of data

TYPE:
Journal Article

FREQUENCY OF UPDATE:
None

NUMBER AND TYPE OF RECORDS: MEAN - time - between - maintenance action for five classes of equipment

DATA BOUNDARY: Ethylene plant pumps; ethylbenzene - styrene monomer plant gas compressors, screw conveyors, pumps, and other non-moving equipment

DATA ACCESS:

Contact: Published in "Reliability Engineering", Vol. 5, 1983, pp.197-215, Elsevier Applied Science Publishers Ltd., Great Britain
Report accessibility: Open Literature/Libraries

DESCRIPTION:

Data from an existing collection system were analyzed for failure modes and distribution. The results of Pareto analyses indicate the principal causes of failure. A few values of mean times to maintenance action (MTBM) are given for ethylene plant pumps (85 electric driven centrifugal pumps over a 19-month period), and ethylbenzene-styrene monomer plant equipment from 10 months data: 4 gas compressors, 3 screw conveyors, 121 pumps, and 235 other items in service. From the Pareto analyses, MTBM values are given according to failure cause for each class of major rotating equipment. In addition, the failure distribution according to process service is given for the pumps.

The article points out the shortcomings of the maintenance system and some steps to improve it.

PROCESS EQUIPMENT DATA SOURCES

TITLE:
Causes of Ammonia Plant Shutdowns: Survey V

SPONSOR/AUTHOR:
G. P. Williams, W. W. Hoehing and R. G. Byington

NO.: 4.4-6

INDUSTRY:
Chemical Process

TIME FRAME:
1965 to 1984

TYPE:
Report

FREQUENCY OF UPDATE:
None

NUMBER AND TYPE OF RECORDS: Records of 5884 shutdowns over 98 years operating time

DATA BOUNDARY: Ammonia Plant

DATA ACCESS:

Contact: Engineering Society Library
American Institute of Chemical Engineers
345 East 47th Street
New York, NY 10017

DESCRIPTION:

This report addresses ammonia plant shutdowns over the listed time period in 40 countries. It provides a basis for comparing plant performance area by area leading to better control of reliability efforts while reducing maintenance and unplanned shutdown costs. Data are presented for shutdowns due to power, equipment, instrumentation, feedstock and product inventory control.

PROCESS EQUIPMENT DATA SOURCES

TITLE:
Reliability Data Collection and Use in Risk and Availability Assessment

SPONSOR/AUTHOR: European Reliability Databank Association (EuReDatA)

NO.: 4.4-7

INDUSTRY:
Varied

TIME FRAME:
Through 1985

TYPE:
Book

FREQUENCY OF UPDATE:
None

NUMBER AND TYPE OF RECORDS: 72 papers, several of which contain some data.

DATA BOUNDARY: Wide variety of systems and components

DATA ACCESS:

Contact: Dr. Hans Jorg Wingender, Editor
NUKEM GmbH, Rodenbacher Chaussee 6
P. O. Box 110080
6450 Hanau 11, FRG

Publisher: Springer - Verlag Berlin, Heidelberg 1986

DESCRIPTION:

This book features the Proceedings of the 5th EuReDatA conference held in Heidelberg, Germany, April 9-11, 1986. It contains 72 papers, several of which contain some data on a wide variety of systems and components. The papers are categorized as follows: 1) Overviews, 2) Reliability Data Banks, 3) Reliability Data Processing, 4) Safety and Reliability Assessment, 5) Data and Uncertainties, 6) Human Reliability, 7) Reliability Modelling and Techniques, 8) Reliability Feedback in Systems Design and Operation, 9) Intelligent Interfaces for Data Retrieval.

PROCESS EQUIPMENT DATA SOURCES

TITLE: Pipeline Reliability: An Investigation of Pipeline Failure Characteristics and Analysis of Pipeline Failure Rates

SPONSOR/AUTHOR: Terje Andersen and Asbjorn Misund Det norske Veritas

NO.: 4.4-8

INDUSTRY: Petroleum and Natural Gas

TIME FRAME: 1966 to 1981

TYPE: Journal Article

FREQUENCY OF UPDATE: None

NUMBER AND TYPE OF RECORDS: Data on frequency and cause of pipeline failures

DATA BOUNDARY: Data is specific to submarine and cross-country oil and natural gas pipelines

DATA ACCESS:

Contact: Journal of Petroleum Technology, Vol. 35, No. 4, April 1983, pgs. 709-717

DESCRIPTION:

This article presents an overview of the causes and frequency of failures for submarine and cross-country pipelines handling oil and natural gas. It gives several tables and charts which include information on the type of pipeline, the cause of the failure, and the number of failures. Data from failures in the US and the North Sea are included. Failure rates based on the total length of piping are calculated.

PROCESS EQUIPMENT DATA SOURCES

TITLE: Fault Tree Analysis Report for Coal-Gasification Process-Development Unit.

SPONSOR/AUTHOR: Powers, G. J.; Lapp, S. A.
Design Sciences INC for Department of Energy

NO.: 4.4-9

INDUSTRY: Coal Gasification

TIME FRAME: June 1982

TYPE: Report

FREQUENCY OF UPDATE: None

NUMBER AND TYPE OF RECORDS: Failure rates (per year basis) for over 400 events from fault trees; Unavailability data

DATA BOUNDARY: Coal-gasification Process Development unit systems and components

DATA ACCESS:

Contact: NTIS, Springfield, VA 22161
Report No. DE82020754
Phone: (703) 487-4650
Report cost: $25.00

DESCRIPTION:

In this study detailed fault trees with probability and failure rate calculations were generated for the events: (1) Fatality due to Explosion, Fire, Toxic Release or Asphyxiation at the Process Development Unit (PDU) Coal Gasification Process; and (2) Loss of Availability of the PDU. The fault trees for the PDU were synthesized by Design Sciences, Inc., and then subjected to multiple reviews by Combustion Engineering. The steps involved in hazard identification and evaluation, fault tree generation, probability assessment, and design alteration are presented in the main body of this report. The fault trees, cut sets, failure rate data and unavailability calculations are included as attachments to this report. Although both safety and reliability trees have been constructed for the PDU, the verification and analysis of these trees were not completed as a result of the curtailment of the demonstration plant project. Certain items not completed for the PDU risk and reliability assessment are listed.

PROCESS EQUIPMENT DATA SOURCES

TITLE: Data Base Development and Equipment Reliability for Phase 1 of the Probabilistic Risk Analysis DPST-87-642

SPONSOR/AUTHOR: Technical Division Savannah River Labs

NO.: 4.4-10

INDUSTRY: Chemical Process, Nuclear Fuel Cycle

TIME FRAME: 1970 to mid 1985

TYPE: Report

FREQUENCY OF UPDATE: Every 5 years

NUMBER AND TYPE OF RECORDS: About 250 component failure rates and 95% upper bounds

DATA BOUNDARY: Pumps, valves, pipe, motors, diesels, heat exchangers, relays, fans for systems given below.

DATA ACCESS:

Contact: D. S. Cramer, Savannah River Plant, Aiken, SC 29868
Phone: (803) 725-1491
Ordering Address: National Technical Information Service (NTIS)
Springfield, VA 22161
(703) 487-4650
Cost: $20 - $50 paper copy, $7 microfiche copy
Accessibility: No restrictions

DESCRIPTION:

The failure events were obtained from Savannah River Plant sources such as Reactor Incident Reports, daily logs, and operating summaries. The records include over 4,400 events going back to 1970 for an effective average of 3 operating nuclear reactors. Some entries represent data averaging about 110 reactor years experience accumulated since operation began.

Systems covered include primary, secondary and emergency core cooling systems, control rod, refueling, scram, electrical power, and lube oil, airsupply, battery supplies.

Components include items like pumps, pipes, valves, motors, gear trains, heat exchangers, relays, switches, fans, thermocouples, and diesel generators.

Some general types of Human Errors are also included.

PROCESS EQUIPMENT DATA SOURCES

TITLE:
Reliability Analysis of Pumps for Uranium Solutions

SPONSOR/AUTHOR:
J. A. Hoffmeister

NO.:
4.4-11

INDUSTRY:
Chemical Process, Nuclear

TIME FRAME:
Unknown (prior to 1987)

TYPE:
Journal Article

FREQUENCY OF UPDATE:
None

NUMBER AND TYPE OF RECORDS: Operating life repair frequency & cost data

DATA BOUNDARY: Dual-diaphragm pumps in uranium solution service.

DATA ACCESS:

Contact: J. A. Hoffmeister, PE
Building 9106, Martin Marietta Energy Systems
P.O. Box 2009, Oak Ridge, TN 37831-8024

Bibliographic Data: IEEE Transactions on Reliability, Vol. 37, No.2
June 1988 (p.144-148)

DESCRIPTION:

This paper describes a reliability analysis of dual - diaphragm pumps in uranium solution service. It is part of the output from a failure modes and effects analysis of the design for a system to be installed at the Oak Ridge Y-12 plant. The study involved collecting data on pumps with Viton and Teflon diaphragms at 10 gpm and 15 gpm.

PROCESS EQUIPMENT DATA SOURCES

TITLE: Emergency Generators: A Reliability Study Based on an Analysis of Failures

SPONSOR/AUTHOR: Ray Stevens
Hartford Steam Boiler Inspection and Insurance Co.

NO.: 4.4-12

INDUSTRY: Chemical Process

TIME FRAME: 1977-1982

TYPE: Journal Article

FREQUENCY OF UPDATE: None

NUMBER AND TYPE OF RECORDS: 138 Emergency Generator Failures

DATA BOUNDARY: Diesel engine driven emergency generators

DATA ACCESS:

Contact: Ray Stevens
Hartford Steam Boiler Inspection and Insurance Co.
Hartford, CT
Bibliographic Data: "Plant Operations and Progress"
Volume 2 No.4, October 1983

DESCRIPTION:

Engine failures over a five year span for 138 emergency generators are listed. Discussion of each type of failure is presented. Data listed includes percentage of failure over the population and time period. All failures were presumably insured by Hartford Steam Boiler and subject to functional inspections and audits.

PROCESS EQUIPMENT DATA SOURCES

TITLE:
Reliability of a Solids-Fluid Handling Process Plant

SPONSOR/AUTHOR:
J. Notman and A.M. Gerrard

NO.:
4.4-13

INDUSTRY:
Ore Processing

TIME FRAME:
Unknown

TYPE:
Journal Article

FREQUENCY OF UPDATE:
N/A

NUMBER AND TYPE OF RECORDS: Failure modes and failure rates covering 829 failures/repairs.

DATA BOUNDARY: Solids handling equipment for one plant: rotary kiln, leaching tank, screwfeeder, and associated items.

DATA ACCESS:

Bibliographic Data: Published in "Reliability Engineering", Vol. 12, 1985, pp. 35-41
Published by Elsevier Applied Science Publishers Ltd.

DESCRIPTION:

Repair and maintenance records were analyzed to determine failure rates and distribution of failure modes. Preliminary findings are reported which include the Weibull distribution characteristics. Failure mode distributions are approximate. Overall mean-time-between-failure is given for the kiln, leach tank, screwfeeder, tank pump, tank gearbox, and kiln gearbox. The study was confined to an analysis of unscheduled repairs and failures.

PROCESS EQUIPMENT DATA SOURCES

TITLE:
Reliability Assessment of Safety/Relief Valves

SPONSOR/AUTHOR:
Aird, R.J.

NO.:
4.4-14

INDUSTRY:
Chemical Process

TIME FRAME:
2 years

TYPE:
Journal Article

FREQUENCY OF UPDATE:
N/A

NUMBER AND TYPE OF RECORDS: 866 Records

DATA BOUNDARY: Safety and Relief Valves

DATA ACCESS:

Bibliographic Data: R.J. Aird
"Reliability Assessment of Safety/Relief Valves"
Trans. Institute of Chemical Engineers,
Vol 60, 1982, pp 314-318

DESCRIPTION:

866 safety relief valve test records were analyzed for 10% deviation from set point.

4.5 INDEX OF CHEMICAL PROCESS QUANTITATIVE RISK ASSESSMENTS

NO.	TITLE	INDUSTRY	NO. & TYPE OF RECORDS	DATA BOUNDARY	PAGE
4.5-1	Hazardous Waste Tank Failure	Chemical Process	Failure rates from published sources or expert judgment	Storage tanks	57.
4.5-2	Risk Analysis of Six Potentially Hazardous Industrial Objects in the Rijnmond Area; A Pilot Study	Chemical Process	Limited failure data from one plant and published data source failure rates	Pumps and piping, valves, measuring devices, controllers and transmitters, electrical, and vessel data	58.
4.5-3	CANVEY: An Investigation of Potential Hazards from Operations in the Canvey Island/Thurrock Area	Chemical Process	Event frequencies estimated from historical data; failure rates from System Reliability Service	Event and equipment failure data for analysis of nine industrial plants and hazardous material transport	59.

CHEMICAL PROCESS QRAS

TITLE:
Hazardous Waste Tank Failure

SPONSOR/AUTHOR: US EPA

NO.: 4.5-1

INDUSTRY: Chemical Process

TIME FRAME: Through 1985

TYPE: Report

FREQUENCY OF UPDATE: None

NUMBER AND TYPE OF RECORDS: Failure rates from published sources or expert judgment

DATA BOUNDARY: Storage tanks

DATA ACCESS:

Contact: National Technical Information Service (NTIS)
Springfield, VA 22161
Phone: (703) 487-4650
Report No.: PB86-192945
Cost: $50.95

DESCRIPTION:

This report documents an extensive study of the likelihood of failure and release of hazardous materials from storage tanks. The report includes a fault tree analysis of a "generic" tank and the quantification of the probability of a release. A lengthy computer program (232 pages of Fortran code) for simulation of incidents is included in the report. In term of reliability data, the report includes a listing of failure probabilities used in the analysis in Table 13. The data sources used and the means by which the data in Table 13 were derived are detailed in Appendix A. The report does not include new, experience based data. Most of the data was taken from other published sources. The gaps were filled in with "engineering estimates" and "personal communications" with various groups such as the Portland Cement Association. This detailed PRA may be of use to the CPI not so much because of the data, but because of its extensive treatment of vessel and pipe leak and rupture probabilities.

CHEMICAL PROCESS QRAS

TITLE: Risk Analysis of Six Potentially Hazardous Industrial Objects in the Rijnmond Area; A Pilot Study

SPONSOR/AUTHOR:	**NO.:**
Dutch Labor Directorate	4.5-2

INDUSTRY:	**TIME FRAME:**
Chemical Process	Through 1981
TYPE:	**FREQUENCY OF UPDATE:**
Book	None

NUMBER AND TYPE OF RECORDS: Limited failure data from one plant and published data source failure rates

DATA BOUNDARY: Pumps and piping, valves, measuring devices, controllers and transmitters, electrical, and vessel data

DATA ACCESS:

Report Title: Risk Analysis of Six Potentially Hazardous Industrial Objects in the Rijnmond Area; A Pilot Study (1982)
Report No.: ISBN-90-277-1393-6
Ordering Address: Kluwer Academic Publishers
P.O. Box 358, Accord Station, Hingham, MA 02018-0358
Phone: (617) 871-6600, Customer Service Department
Cost: $165; no postage charge if pre-paid order

DESCRIPTION:

The Rijnmond area is that part of the Rhine delta between Rotterdam and the North Sea. The Commission for the Safety of the Population at large (COVO) commissioned the study for six chemicals and the operations associated with them: acrylonitrile, liquid ammonia, liquid chlorine, LNG, propylene, and part of a separation process (diethanolamine stripper of a hydrodesulfurizer). The study objectives were to evaluate methods of risk assessment and obtain experience with practical applications of these methods. The results were to be used to decide to what extent such methods can be used in formulating safety policy. The study was not concerned with the acceptability of risk or the acceptability of risk reducing measures.

The book contains only a small amount of actual failure data from one of the study plants. One year's worth of data is described for acrylonitrile storage tank instrumentation (level and temperature), plus some assessment of hose failures. The remaining base failure rate data are contained in Appendix IX of the report with references to the sources of data, many of which are commonly cited data sources. The data are classified under Pumps and Piping, Valves, Measuring Devices, Controllers and Transmitters, General Electrical, and Vessels. Correction factors were applied to these base data before using in fault trees, etc, to account for environment, maintenance, and operating philosophy as judged by the risk assessment team.

CHEMICAL PROCESS QRAS

TITLE: CANVEY: An Investigation of Potential Hazards from Operations in the Canvey Island/Thurrock Area

SPONSOR/AUTHOR:	**NO.:**
UK Health & Safety Commission	4.5-3

INDUSTRY:	**TIME FRAME:**
Chemical Process	Through 1977

TYPE:	**FREQUENCY OF UPDATE:**
Report	None

NUMBER AND TYPE OF RECORDS: Event frequencies estimated from historical data; failure rates from System Reliability Service.

DATA BOUNDARY: Event and equipment failure data for analysis of nine industrial plants and hazardous material transport.

DATA ACCESS:

Report Title: CANVEY: An Investigation of Potential Hazards from Operations in the Canvey Island/Thurrock Area (Essex, U.K)

Ordering Address: H.M. Stationery Office
P.O. Box 569
London, England SE1 9NH

DESCRIPTION:

This study investigated risks to the public from serious accidents which could occur at the industrial facilities in this part of Essex, U.K. Results are expressed as risk to an individual and societal risk from both existing and proposed installations. Risk indices were also determined for modified versions of the facilities to quantify the risk reduction from recommendations in the report. Nine industrial plants were analyzed along with hazardous material transport by water, road, rail and pipeline. The potential toxic, fire and explosion hazards were assessed for flammable liquids, ammonia, LPG, LNG, and hydrogen fluoride (HF). The 24 appendices to the report cover various aspects of the risk analysis. These include: causes and effects of unconfined vapor cloud explosions; fire impacts; dispersion of ammonia, HF, and heavy gas; and pressure vessel failure rates.

The report does not contain original data, only references to sources of information. Frequencies were estimated for events from historical data and by incorporating failure rate data from the System Reliability Service. Event frequencies and equipment failure rates in the report are coded to indicate the confidence level of the data. In particular, the coding reflects whether the data is (a) assessed statistically from historical data, (b) based on statistics but reflecting some judgment, (c) estimated by comparison to previous cases for which fault tree analysis have been made, and (d) "dummy" figures – where subjective judgement must be made even though there is likely to be large uncertainly in the values.

4.6 INDEX OF NON-PROCESS EQUIPMENT DATA BASES

NO.	TITLE	INDUSTRY	NO. & TYPE OF RECORDS	DATA BOUNDARY	PAGE
4.6-1	Centralized Reliability Data Organization Input Guide	Nuclear	1800 Component failure events causing abnormal operation; engineering data for 20,000 components found in liquid metal reactors (e.g., LMRs). Gives generic data for 45 equipment items (valves, pumps, etc.)	Liquid metal reactor sites and test facility systems and components	62.
4.6-2	Nuclear Plant Reliability Data System	Nuclear	Engineering data from all U.S. NPP (currently 91 plants); failure data describing 44,000 events	Voluntary reporting of engineering information and failures for selected systems and components as defined in a reportable scope manual	64.
4.6-3	The European Reliability Data System: An Organized Information Exchange on the Operation of European Nuclear Reactors	Nuclear	Comprehensive records on equipment failure,frequency, modes, unusual events, and plant production	U.S. & European nuclear reactor data on equipment performance, repair and maintenance	65.
4.6-4	Determination of Reliability Characteristic Factors in the Nuclear Power Plant Biblis B, Gesellschaft fur Reaktorsicherheit mbH	Nuclear	Failure rates with upper and lower bounds and maintenance data for 17,000 components from 37 safety systems	Data for pumps, valves, and electrical positioning devices, electric motors and drives from an operating power plant	66.
4.6-5	Generating Availiability Data System	Power	System and equipment failure data from 2,600 electric power units giving failure rates, modes, mean time to repair, mean time between outages	All types of major electrical power generating equipment which could cause a full or partial outage of an electrical generating utility	68.
4.6-6	Reliability Data Book for Components in Swedish Nuclear Power Plants	Power, Nuclear	30,000+ recorded events	Safety and commercial grade components, i.e. pumps, valves, diesels, filters, tanks, and heat exchangers from 4 nuclear and non nuclear power generating plants	70.
4.6-7	Failure and Inventory Reporting System	Offshore Oil and Natural Gas	8,000 failure events and causes; Inventory data on ASME coded devices	Safety and pollution prevention devices on offshore structures, e.g. subsurface safety valves	72.
4.6-8	Government Industry Data Exchange Program (GIDEP)	Varied	Recorded historical, experimental, and test data from 650 participating agencies/industries. May include failure frequencies, rates, modes, mechanisms as well as repair times	Engineering, Reliability, Metrology, and Failure data on mechanical, electronic and electrical components	73.
4.6-9	System Reliability Service	Varied	Status reports contain failure, repair, and maintenance data on 450 categories; data sets are broader	A wide variety of equipment failure and performance records from utilities and manufacturers, plus collected inspection data	75.
4.6-10	SAIC Data Base	Nuclear	Plant maintenance and repair records by system and component type; selected basic event failure rates and unavailability	Data mainly for pumps, valves, diesels, batteries, chargers, and heat exchangers	76.

4.6 INDEX OF NON-PROCESS EQUIPMENT DATA BASES

NO.	TITLE	INDUSTRY	NO. & TYPE OF RECORDS	DATA BOUNDARY	PAGE
4.6-11	The In-Plant Reliability Data Base for Nuclear Plant Components	Nuclear	Corrective maintenance records: 4000 for pumps; 5812 for valves; 698 for electrical failure rates and modes.	Pumps, valves, diesels, battery chargers, and inverters from 4 nuclear power plants	78.
4.6-12	IEEE Standard 500-1984	Nuclear	Over 100 data sheets, containing failure rate estimates for various failure modes and some repair times.	Electrical, instrumentation, and mechanical components in nuclear power plants	80.
4.6-13	Generic Data Base for Data and Models Chapter of the National Reliability Evaluation Program Guide	Nuclear	Estimates of hourly and per demand failure rates for 20 categories with error factors generated by experts	Standard nuclear plant PRA equipment list	82.
4.6-14	Offshore Reliability Data Handbook	Offshore Oil	Failure rates based on 14,000 failures for 300 component types. Mean values, upper and lower bounds are offered for different failure modes.	Offshore oil equipment data and boundaries for safety systems, process systems, electrical systems, utility systems, cranes and drilling equipment.	84.
4.6-15	RADC Non-Electronic Reliability Notebook	Military	Failure rates based on operations and maintenance records and industry-wide sources	Covers 250 non-electronic part types such as actuators, complings, filters for 12 classes of environment (air, ground, sea) used in military systems and equipment.	87.
4.6-16	Reliability Prediction of Electronic Equipment (Military Handbook 217E)	Government and Military	Failure rates for 13 categories of components.	Electronic components used in military electronic systems and equipment.	89.

NON-PROCESS EQUIPMENT DATA BASES

TITLE:

Centralized Reliability Data Organization Input Guide

SPONSOR/AUTHOR: U.S. DOE and Japan's Power Reactor and Nuclear Fuel Development Corp. (PNC)

NO.: 4.6-1

INDUSTRY: Nuclear

TIME FRAME: 1960 to present

TYPE: Data Base and Report

FREQUENCY OF UPDATE: Quarterly

NUMBER AND TYPE OF RECORDS: 1800 component failure events causing abnormal operation; engineering data for 20,000 components found in liquid metal reactors (e.g., LMRs). Gives generic data for 45 equipment items (valves, pumps, etc.)

DATA BOUNDARY: Liquid metal reactor sites and test facility systems and components

DATA ACCESS:

Contact: Mr. H. E. Knee, Oak Ridge National Laboratory, Bldg. 6205, ms-360
P.O. Box 2008, Oak Ridge, TN 37831-6360
Phone: (615) 574-6163 (FTS 624-6163)
Data accessibility: Limited

DESCRIPTION:

The Centralized Reliability Data Organization (CREDO) is maintained at ORNL to provide a central computer-based source of accurate, timely data and information for use in reliability, availability and maintainability analyses of liquid metal reactors (LMRs). CREDO is a component-based system, that addresses a comprehensive list of 45 generic components that are representative of all components found at LMRs.

The data base management system used for CREDO catalogs and stores data in three types of files that correspond to their respective data types: (a) Engineering Data; data and information on a component's design and operating characteristics, (b) Operating Data; A Chronological sequence of reactor operating experience, per reactor mode, (c) Event Data; data on abnormal operating events that are attributable to a component, including event descriptions, method of detection, failure mode/cause, and corrective action.

The CREDO data base contains data from The Fast Flux Test Facility in Richland, Washington, The Experimental Breeder Reactor - II in Idaho Falls, Idaho, The test loops of the Energy Technology Engineering Center (ETEC) in Canoga Park, California, The JOYO Liquid Metal Fast Breeder Reactor at the O-Arai Engineering Center (OEC) in Japan, and the test loops of OEC.

Data Resource 4.6-1
Centralized Reliability Data Organization Input Guide
Example Data Sheet

EVENT DATA REPORTING FORM (CREDO-1, Rev. 1)

1. REPORT IDENTIFICATION (a) Report I.D. No. ______________
(CREDO Use Only)

(b) Site* YOURSITE (c) Unit* YOURUNIT (d) Report Date 11/18/85
Mo. Da. Yr.

(e) Occurrence Title Failed Flux Monitoring Channel

(f) Report Status: ☒ New ☐ Chng (g) Previous Report Date __/__/__
Mo. Da. Yr.

(h) Related Reports: UOR YOUR-85-11 Other ______________

(i) Event Date 11/03/85 Event Time ____hrs.
Mo. Da. Yr.

2. EVENT NARRATIVE Reactor operator had last monitored widerange channel C at 0730, whereby it indicated as the others. When it was agin selected at 0745 it read ≈20% lower than channels A&B. He rechecked A&B widerange channels,
(Attach additional narrative on separate sheet as desired)

3. EVENT DETECTION/IMMEDIATE ACTION (a) Detection Date 11/03/85 Time 0245 hrs.
Mo. Da. Yr.

(b) Method of Detection ROUTINE_ (c) Time/Detection to Initial Action* __0_____

(d) Operating Status: Unit* PWROPS__ System* NORMOPS_
Subsystem* MAXNORM_

(e) Initial (Immediate) Action Monitored All other flux monitoring channels to compare indications. When it was decided

4. COMPONENT FAILURE DATA

(a) Component Name* SIGMOD__ (b) CREDO I.D. No.* SMO2OLOO

(c) Site I.D. No.* 16-NAM-13A (d) System* FLUXMON_ (e) Subsystem* WRFLXMON

(f) Component Description Solid state linear voltage amplifier

(g) Failure Type* ELECTRON (h) Failure Mode* ABNORMOP

(i) Failure Cause* OPENCKT_ (j) ☒ Primary ☐ Secondary

(k) Failure Cause Narrative A resistor (R-2) in the input circuitry opened changing the input impedence, which resulted in the reduced output. The reason for the resistor failure is not known, however it probably just reached end of life.
(Attach additional narrative on separate sheet as desired)

(l) Failure Effects: System Reduced – only 2 WR-channels (Hours Lost = 0)

Unit NONE (Hours Lost = 0)

Other Items Affected Plant protection: SCRAM

(m) Critical Parts resistor

NON-PROCESS EQUIPMENT DATA BASES

TITLE:
Nuclear Plant Reliability Data System

SPONSOR/AUTHOR:
Institute of Nuclear Power Operations (INPO)

NO.:
4.6-2

INDUSTRY:
Nuclear

TIME FRAME:
1974 to Present

TYPE:
Data Base and Reports

FREQUENCY OF UPDATE:
Continuous

NUMBER AND TYPE OF RECORDS: Engineering data from all U.S. NPP (currently 91 plants); failure data describing 44,000 events

DATA BOUNDARY: Voluntary reporting of engineering information and failures for selected systems and components as defined in a reportable scope manual

DATA ACCESS:
Contact: Keith Pope, INPO, 1100 Cir. 75 Pkwy, Ste 1500, Atlanta, GA 30339
Phone: (404) 953-5443
Report ordering address: Same as above, For annual summaries of data base for 1978-1980: GPO Sales Program, Div. of TIDC, U.S. NRC, Washington, D.C. 20555
Report Cost: Free to INPO members/participants;
NRC access is $145 per connect hour
Report accessibility: Summary reports available; Limited specific reports.
Computer access: Limited, as with report access

DESCRIPTION:

The NPRDS is an industry-wide system for monitoring the performance of selected systems and components at U.S. commercial nuclear power plants. Information in NPRDS is derived from a standardized format input report prepared by U.S. nuclear plant licensees. The plants are asked to submit failure reports on catastrophic events and degraded failures within the defined reportable scope; reporting of incipient events is optional. Command faults are not reportable unless they make an entire system unavailable. In addition, the plants are asked to file component engineering reports on all components within the selected systems and reportable scope. These reports contain detailed design data, operating characteristics, and performance data on the selected systems and components (over 3000 components, from approximately 30 systems, per unit). The selected systems are primarily safety systems.

NPRDS data are available to users, either through annual summary reports (to resume publication in the coming year) or through direct on-line data base access from a computer terminal. Special reports and listings are available through specific requests for extraction or data analysis.

Although the reporting period covers approximately 12 years, the most detailed reporting has been implemented by most of the licensees since 1984 when the new LER rule came into effect and fewer component failures were reported to the LER system. One should note that no human-error-related data are included in the data bank except for actions that cause broken hardware.

NPRDS data can be obtained in the form of magnetic tapes or floppy disks that are dBASE III or LOTUS 1-2-3 compatible.

NON-PROCESS EQUIPMENT DATA BASES

TITLE: The European Reliability Data System: An Organized Information Exchange on the Operation of European Nuclear Reactors

SPONSOR/AUTHOR:	**NO.:**
European Economic Community	4.6-3

INDUSTRY:	**TIME FRAME:**
Nuclear	Early 1970's to Present
TYPE:	**FREQUENCY OF UPDATE:**
Data Base	Every 6 months

NUMBER AND TYPE OF RECORDS: Comprehensive records on equipment failure, frequency, modes, unusual events, and plant production

DATA BOUNDARY: U.S. & European nuclear reactor data on equipment performance, repair and maintenance

DATA ACCESS:

Contact: G. Mancini
Sector Head-Nuclear Installations Engineering Division
Ispra Establishment
I-21020 Ispra (Varese)
Italy

Phone: (0332) 789714
(Telex: 380042 EURI)

DESCRIPTION:

The ERDS is comprised of four data banks:

1. The Component Event Data Bank contains raw data on component failure, reliability, and operating modes, from EEC countries, Spain, and Sweden.

2. The Abnormal Occurrences Reporting System contains unusual plant events and human factors associated with normal operation from the U.S., Sweden, and all relevant ECC countries except Germany (discussions are in progress there).

3. The Operating Unit Status Report contains production and outage log information on power reactor operation from ECC countries, Spain, and Switzerland.

4. The Generic Reliability Parameter Data Bank is a bibliography of reliability information for plants with similar classes of components.

The ERDS attempts to establish a uniform method of encoding the types of data submitted from the various entities. Inconsistencies of detail are monitored automatically by the data bank computer services.

The data are very comprehensive with direct applications to reliability, risk, and event analysis of nuclear power plants. Information has been assembled on failure frequency, modes, repairs, and maintenance. Rate information is based on demands calculated. The time period covered varies from the early 1970's to the present. Using real time access, the output format if the event can be varied by selection of 20 generic and detailed categories.

NON-PROCESS EQUIPMENT DATA BASES

TITLE: Determination of Reliability Characteristic Factors in the Nuclear Power Plant Biblis B, Gesellschaft fur Reaktorsicherheit mbH

SPONSOR/AUTHOR: RWE

NO.: 4.6-4

INDUSTRY: Nuclear

TIME FRAME: 1977 to Present

TYPE: Data Base

FREQUENCY OF UPDATE: Continuous

NUMBER AND TYPE OF RECORDS: Failure rates with upper and lower bounds and maintenance data for 17,000 components from 37 safety systems.

DATA BOUNDARY: Data for pumps, valves, electrical positioning devices, electric motors and drives from an operating power plant

DATA ACCESS:

Contact: P. Homke
Gesellschaft fur Reaktorsicherheit mbH
Glockengasse 2.5000
Koln 1, West Germany (FRG)
Phone: (02-21) 20 68-1
Report Cost: Free to authorized users of the data base

DESCRIPTION:

The German Gesellschaft fur Reaktorsicherheit (GRS) has a private arrangement with Rheinische Westalisches Elekrizitatswerke (RWE) to compile reliability data from an operating power plant, Biblis B. The data base contains failure rate, maintenance, and operational event data. External event data (floods, earthquake, fire, etc.) are compiled through a separate utility-sponsored data base. The data base provides information on repair and maintenance, and equipment performance.

This data collection effort was concentrated on the following components because of their extensive populations and repair action documentation: pumps, valves, electrical positioning devices, electric motors, and drives. For each component type, preface pages and data summary tables are provided. Separate data summary tables are provided for each component type and are structured in a format that allows for the inclusion of the number of pieces of operating equipment, the total number of operating hours, total number of failures, and hourly failure rates with upper and lower bounds.

All of the information listed directly above is correlated to an all damages mode, individual subcategories of failure modes (internal leakage, for instance), and the influence parameters and their subdivisions. Real-time access is possible through a data base management system.

Data Resource 4.6-4

Determination of Reliability Characteristic Factors in the Nuclear Power Plant Biblis

Example Data Sheet

TABLE: SUMMARY OF
T 11 FAILURE RATES FOR: regulation damper (KAF), controlled SHEET: 1

description	No. of pieces of oper. equipt.	sum of operating hours	No. of failures	λ_{iu} $\frac{10^{-6}}{h}$	$\hat{\lambda}$ $\frac{10^{-6}}{h}$	λ_{io} $\frac{10^{-6}}{h}$
damage type: all damages	169	2,535,000	30	8.5	11.8	15.9
all damages with non-availability during repair AM ≥ 03	169	2,535,000	25	6.8	9.8	13.7
failure mode: does not open and/or close						
BM - type average	169	2,535,000	25	6.8	9.8	13.7
influence parameter:						
system (reactor ventilation) TL	21	315,000	9	14.9	28.6	49.8
(air-cond. and ventilation) UV	148	2,220,000	16	4.5	7.2	11
utilization mode: (continuous operation)	94	1,410,000	14	6	9.9	15.4
(cyclic operation)	5	75,000	4	18.1	53.3	122
(stand-by operation)	70	1,050,000	7	3.2	6.7	12.6

NON-PROCESS EQUIPMENT DATA BASES

TITLE:

Generating Availability Data System

SPONSOR/AUTHOR:	**NO.:**
NERC	4.6-5

INDUSTRY:	**TIME FRAME:**
Power	Mid-1960's to Present

TYPE:	**FREQUENCY OF UPDATE:**
Data Base and Reports	Quarterly

NUMBER AND TYPE OF RECORDS: System and equipment failure data from 2,600 electric power units giving failure rates, modes, mean time to repair, mean time between outages.

DATA BOUNDARY: All types of major electrical power generating equipment which could cause a full or partial outage of an electrical generating utility

DATA ACCESS:

Contact: Ronald Niebo, Director
North American Electric Reliability Council
101 College Road East, Princeton, NJ 08540-6601

Phone: (609) 452-8060

Report Cost: Free to GADS Participants

Data accessibility: General information available as annual or quarterly status reports

DESCRIPTION:

The NERC GADS encompasses 2,600 electric power units, including nuclear, fossil, hydro, and combined cycle. The data base is comprised of safety or commercial quality components, 25 percent electronic and 75 percent mechanical. There are over a million individual and statistically reduced events, encoded in a structured extended text format. Historically, system transients and external initiating events have been grouped together. Beginning in 1982, the categories started being reported separately. Data supplied to GADS have been recorded by the electric utilities participating in the GADS program. All generic categories of components are included if their failure caused a forced, scheduled, or partial plant outage. The data are statistically reduced to derive failure frequency rates, modes, mechanisms, and intervals using calculated population and demands. Also considered are the mean repair time and equipment downtime.

The NERC GADS ten-year review report for 1971-1980 on equipment availability presents statistical data sets on the performance of major types of electrical power generating units. Cumulative and unit-year averages are calculated on such quantities as service hours, available hours, scheduled outage rate, mean time between full forced outages, shutdown because of economic reasons, and probability of outage. The number of start demands and successful starts are included.

The ten-year review for 1971-1980 of the NERC GADS component cause code report documents all outages attributed to each of the component cause codes.

Data Resource 4.6-5
Generating Availability Data System
Example Data Sheet

Fossil Units - All Fuel Types
All Size Ranges 1983-1987

LOSSES DUE TO OUTAGES AND DERATINGS
BY CAUSE CODE CATEGORIES

CAUSE CODE RANGES	SYSTEM/COMPONENT CAUSE CODE CATEGORIES	FORCED OUTAGES			FORCED DERATINGS			FORCED AND SCHEDULED OUTAGES AND DERATINGS		
		AVG NO OCC PER UNIT-YR	AVG MWH PER UNIT-YR	AVG HRS PER UNIT-YR	AVG NO OCC PER UNIT-YR	AVG EQV MWH PER UNIT-YR	AVG EQV HRS PER UNIT-YR	AVG NO OCC PER UNIT-YR	AVG EQV MWH PER UNIT-YR	AVG EQV HRS PER UNIT-YR
0010-1999	BOILER	4.84	70232	212.24	6.93	27187	81.91	2.05	265355	833.18
0010-0480	.Boiler Fuel Supply	0.46	2302	7.30	11.84	7444	24.58	13.93	14794	50.90
0010-0110	..Coal Hand. Equip. up Thru Bunkers	0.02	506	1.03	0.77	657	1.53	0.83	1314	3.27
0200-0350	..Pulv., Prim. Air Fans & Assoc. Duct	0.10	525	1.35	9.94	5962	20.22	11.47	8803	29.38
0360-0410	..Burners	0.23	874	2.88	0.56	347	1.50	0.88	3243	12.47
0415-0435	..Cyclone	0.02	109	0.49	0.40	159	0.53	0.45	580	2.44
0440-0480	..Oil and Gas Sys. (except light-off)	0.10	289	1.56	0.17	320	0.79	0.30	853	3.34
0500-0799	.Boiler Piping System	0.33	5010	12.84	0.81	1781	5.30	1.40	12817	38.57
0500-0530	..Main Steam	0.08	1584	4.22	0.14	231	1.05	0.32	3465	12.07
0540-0570	..Cold and Hot Reheat Steam	0.03	1085	2.20	0.21	585	1.29	0.26	3170	8.28
0580-0620	..Desuperheaters/Attemperatures	0.03	271	0.89	0.14	132	0.36	0.20	700	2.26
0630-0660	..Startup Bypass	0.05	734	1.27	0.03	83	0.13	0.09	1402	2.44
0670-0730	..Feedwater and Blowdown	0.04	314	1.06	0.04	42	0.19	0.11	706	2.56
0740-0770	..Boiler Recirculation	0.07	622	1.70	0.24	684	2.20	0.36	1787	5.24
0775-0799	..Miscellaneous Piping	0.03	400	1.50	0.01	23	0.09	0.06	1586	5.72
0800-0859	.Boiler Internals and Structures	0.10	1691	5.05	0.15	397	1.95	0.36	4256	15.76
0860-0920	.Slag and Ash Removal	0.05	684	1.99	0.61	543	1.37	0.79	2111	6.04
1000-1090	.Boiler Tube Leaks	2.76	50955	156.82	0.60	1143	4.42	3.88	61673	196.87
1100-1200	.Tube Fireside Slagging or Fouling	0.08	1754	5.23	1.05	1715	4.43	1.27	5268	14.60
1300-1350	.Miscellaneous Boiler Tube Problems	0.02	380	1.25	0.06	315	1.10	0.15	7513	25.54
1400-1599	.Boiler Air and Gas Systems	0.38	3864	11.01	2.71	9956	26.93	3.56	22738	66.64
1400-1450	..Air Supply	0.15	1446	3.71	0.82	3563	10.15	1.11	6204	19.05
1455-1530	..Flue Gas	0.22	2266	6.67	1.71	6098	16.13	2.22	15083	42.74
1535-1580	..Flue Gas Recirculation	0.01	87	0.41	0.04	190	0.45	0.06	581	1.94
1590-1599	..Miscellaneous (Air Supply)	0.01	64	0.22	0.14	104	0.20	0.17	870	2.92
1700-1799	.Boiler Control Systems	0.55	2433	6.55	0.63	472	1.30	1.28	4469	16.65
1800-1820	.Boiler Overhaul and Inspections	0.01	434	2:11	0.05	131	0.37	0.73	121418	373.27
1850-1850	.Boiler Water Condition	0.03	159	0.61	2.18	2451	7.40	2.35	2822	8.69
1900-1910	.Boiler Design Limitations	0.00	20	0.08	0.05	353	0.69	0.06	546	1.26
1980-1999	.Miscellaneous (Boiler)	0.08	547	1.41	0.40	485	2.05	0.69	4931	18.39
3110-3999	BALANCE OF PLANT	1.39	14248	40.75	8.27	19823	58.15	12.64	52627	165.61
3110-3199	.Condensing System	0.25	1813	5.66	2.34	1905	6.12	4.24	7152	25.35
3110-3119	..Condenser Tubes	0.10	908	3.01	1.90	1489	4.68	3.55	4553	15.50
3120-3129	..Condensing Casing/Shell & Internals	0.02	526	1.22	0.05	85	0.24	0.10	963	2.7.
3130-3149	..Vacuum Equipment	0.11	252	0.95	0.26	203	0.64	0.38	499	1.85
3150-3159	..Condenser Controls	0.01	18	0.07	0.01	13	0.26	0.03	38	0.37
3170-3199	..Miscellaneous (Condensing System)	0.01	108	0.41	0.12	114	0.30	0.19	1099	4.92
3210-3299	.Circulating Water System	0.08	860	4.00	0.73	1692	5.86	1.01	4531	20.37
3310-3399	.Condensate System	0.11	902	3.03	0.94	2524	7.27	1.20	5119	18.67
3310-3330	..Pumps, Piping and Valves	0.03	205	0.69	0.44	1753	5.04	0.57	2357	6.88
3340-3344	..Low/Inter. Press. Heater/Deaerators	0.05	508	1.21	0.22	462	1.41	0.33	2118	9.11
3350-3352	..Polishers/Chemical Addition	0.01	99	0.23	0.22	234	0.61	0.24	417	1.07
3360-3399	..Miscellaneous (Condensate System)	0.01	89	0.92	0.05	74	0.21	0.07	227	1.61
3401-3499	.Feedwater System	0.47	4867	12.33	3.33	12196	34.54	4.38	24295	66.31
3501-3509	.Heater Drain Systems	0.01	85	0.20	0.19	222	0.74	0.22	445	1.36
3520-3529	.Extraction Steam	0.02	215	0.49	0.04	140	0.45	0.08	551	1.56
3600-3689	.Electrical	0.37	4952	13.78	0.33	733	2.22	0.86	7527	23.65
3810-3899	.Auxiliary Systems	0.07	434	0.93	0.19	108	0.31	0.32	992	3.01
3810-3819	..Service Water (Open System)	0.01	54	0.16	0.02	12	0.04	0.03	157	0.63
3820-3829	..Closed Cooling Water Systems	0.01	34	0.12	0.02	26	0.07	0.04	166	0.64
3830-3839	..Auxiliary Steam	0.02	239	0.39	0.13	41	0.14	0.17	445	0.93
3840-3849	..Service Air	0.00	7	0.03	0.01	13	0.02	0.01	24	0.07
3850-3859	..Instrument Air	0.02	53	0.14	0.02	10	0.02	0.04	90	0.39
3860-3869	..Fire Protection System	0.00	12	0.03	0.00	2	0.00	0.01	19	0.05
3899-3899	..Miscellaneous (Auxiliary Systems)	0.00	35	0.07	0.00	4	0.01	0.01	92	0.30
3950-3999	.Miscellaneous (Balance of Plant)	0.02	121	0.33	0.19	303	0.64	0.33	2016	5.34
4000-4499	STEAM TURBINE	1.22	21628	62.23	1.81	4474	12.64	4.46	114124	378.01
4000-4099	.High Pressure Turbine	0.09	4159	12.51	0.18	1510	5.00	0.34	9864	36.86
4100-4199	.Intermediate Pressure Turbine	0.02	1609	3.10	0.06	165	0.37	0.10	2599	6.00
4200-4250	.Low Pressure Turbine	0.07	4380	11.75	0.25	645	1.23	0.37	9760	26.15
4260-4269	.Valves	0.20	2167	6.62	0.42	868	2.27	0.92	7005	22.48
4270-4279	.Piping	0.03	536	1.50	0.02	30	0.07	0.06	857	2.56
4280-4289	.Lube Oil	0.06	1585	4.27	0.01	20	0.09	0.11	2428	7.21
4290-4309	.Controls	0.42	2186	5.84	0.21	355	0.99	0.77	3920	12.04
4400-4499	.Miscellaneous (Steam Turbine)	0.33	5006	16.62	0.65	881	2.62	1.80	77692	264.72
4500-4899	GENERATORS	0.53	16053	45.34	0.50	1112	2.77	1.35	34400	97.69
4500-4590	.Generator	0.09	6272	21.62	0.09	605	1.36	0.27	12207	37.98
4600-4609	.Exciter	0.14	2747	6.87	0.05	92	0.17	0.23	3458	9.57
4610-4650	.Cooling System	0.09	2928	7.12	0.24	318	0.84	0.39	4999	13.38
4700-4750	.Controls	0.16	753	2.11	0.03	30	0.08	0.24	967	2.89
4800-4899	.Miscellaneous (Generator)	0.06	3353	7.61	0.10	69	0.32	0.23	12770	33.87
8000-8699	POLLUTION CONTROL EQUIPMENT	0.09	1260	3.96	1.47	2435	6.46	1.97	12167	35.65
8000-8499	.Wet Scrubbers	0.02	313	0.78	0.58	1171	2.54	0.75	4480	13.40
8000-8099	..Chemical Supply	0.00	14	0.02	0.02	26	0.06	0.02	41	0.09
8100-8199	..Scrubber	0.00	36	0.13	0.21	450	0.95	0.31	1253	3.01
8200-8299	..Piping, Ducting, Dampers and Fans	0.01	154	0.39	0.21	194	0.49	0.24	842	2.27
8300-8399	..Waste Disposal and Recovery	0.00	24	0.05	0.03	63	0.19	0.04	98	0.27
8400-8499	..Miscellaneous (Wet Scrubbers)	0.01	84	0.19	0.11	439	0.84	0.13	2246	7.76
8500-8549	.Dry Scrubbers	0.00	0	0.00	0.02	15	0.03	0.03	26	0.06
8550-8580	.Precipitators	0.07	915	3.07	0.66	999	3.13	0.98	6834	19.82
8600-8699	.Misc. (Pollution Control Equipment)	0.00	32	0.10	0.20	250	0.76	0.22	827	2.37
9000-9320	EXTERNAL	0.13	1533	3.73	3.74	2445	8.01	3.91	5587	16.44
9000-9040	.Catastrophe	0.03	1109	2.31	0.03	39	0.27	0.06	1166	2.75
9130-9160	.Economic	0.01	107	0.60	0.07	142	0.32	0.09	1631	4.72
9200-9290	.Fuel Quality	0.04	147	0.37	3.51	2106	6.94	3.55	2271	7.39
9300-9320	.Miscellaneous (External Problems)	0.06	170	0.46	0.13	158	0.47	0.21	519	1.58
9500-9720	REGULATORY, SAFETY AND ENVIRONMENTAL	0.01	205	1.65	2.91	2825	10.09	3.01	4077	16.68
9500-9590	.Regulatory	0.00	1	0.00	0.01	21	0.13	0.02	460	2.29
9600-9655	.Stack Emission	0.01	111	0.87	2.61	2464	9.10	2.66	2770	11.08
9660-9696	.Operating Environmental Limitations	0.00	89	0.74	0.28	337	0.85	0.32	723	2.87
9700-9720	.Safety	0.00	3	0.03	0.00	2	0.01	0.01	125	0.44
9900-9920	PERSONNEL ERRORS	0.29	2338	7.10	0.12	130	0.32	0.42	2656	7.91
9999-9999	PERFORMANCE	0.00	25	0.06	0.02	8	0.03	0.08	217	0.76

NON-PROCESS EQUIPMENT DATA BASES

TITLE:
Reliability Data Book for Components in Swedish Nuclear Power Plants

SPONSOR/AUTHOR:	**NO.:**
Swedish State Power Board	4.6-6

INDUSTRY:	**TIME FRAME:**
Power, Nuclear	1973 to Present

TYPE:	**FREQUENCY OF UPDATE:**
Data Base	Quarterly

NUMBER AND TYPE OF RECORDS: 30,000+ recorded events

DATA BOUNDARY: Safety and commercial grade components, i.e. pumps, valves, diesels, filters, tanks, and heat exchangers from 4 nuclear and non nuclear power generating plants.

DATA ACCESS:

Contact: Jean-Pierre Bento
Radet Fur Karnkraft Sakerhet
Nuclear Safety Board of the Swedish Utilities
Box 5864 S-102 48
Stockholm, Sweden

Phone: 046-08-679540
(Telex: 131-08)

DESCRIPTION:

The Swedish Thermal Power Reliability Data System (ATV) is maintained and managed by the Swedish State Power Board at Stockholm, Sweden. Engineering and reliability data have been collected from both nuclear and nonnuclear power generating plants. Nuclear data collection began in 1973. Collection of reliability data began in 1976. Over 30,000 events have been recorded in the data base.

All data recorded in the data base have been acquired from plant records. Statistical reductions of data for generation of reports or specific end use are available. Data are currently collected from four operating plants (eight units). Time clocks have been installed on components to record actual exposure time. Event data are available on a broad variety of safety and commercial grade components including pumps, valves, transformers, diesels, filters, tanks (vessels), and heat exchangers.

Real-time access to the data base is possible through a data base management system. In addition to safety-related system data, data for interfacing (auxiliary) system components are also available. The data base management system provides in-depth flexibility for generation of specific output reports. Although access has been limited to members of the Swedish Utility Consortium and the Swedish Power Board Directorate, a Reliability Data Book is available with generic failure rates based primarily on this data base.

Data Resource 4.6-6

Reliability Data Book for Components in Swedish Nuclear Power Plants

Example Data Sheet

Centrifugal pump, horisontal and vertical

Backgrond data

Number of components:	18
Failure mode:	Failure to start
Number of demands:	784 (per operational time)
Number of failures:	4
Value of Alfa:	0.387
Value of Beta:	75.4
Method of estimation:	Weighted aPriori Moment Method (WPM)

Physical boundary of the component

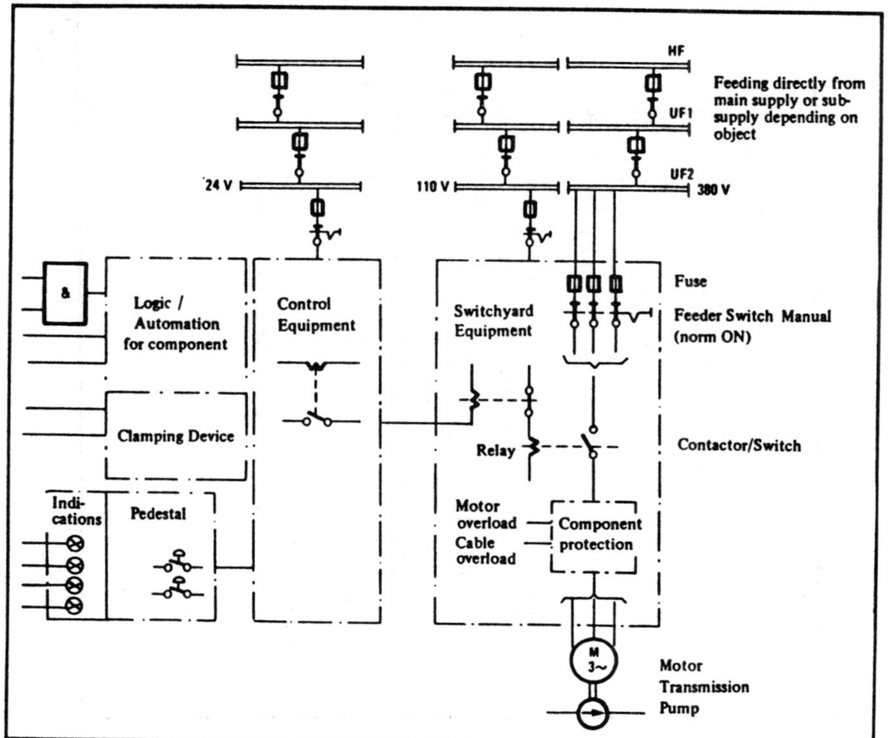

Centrifugal pump, horisontal and vertical

Flow rate:	120 - 240 kg/s
Developed head:	1,2 - 1,8 MPa
Operational mode:	standby

Power plant	Failure to start (10^{-3}/demand)	Active repair (average) (h)
Barsebäck 1	7.0	5
Barsebäck 2	3.3 *	–
Forsmark 1	4.0 *	–
Forsmark 2	4.3 *	–
Oskarshamn 1	11.	2
Oskarshamn 2	2.6 *	–
Ringhals 1	2.8 *	–
All BWR plants		
Mean value	5.1	3
95%	21.	

* No critical failures reported

NON-PROCESS EQUIPMENT DATA BASES

TITLE:

Failure and Inventory Reporting System

SPONSOR/AUTHOR:	**NO.:**
U.S. Dept. of the Interior	4.6-7

INDUSTRY:	**TIME FRAME:**
Offshore Oil and Natural Gas	1980 to Mid-1982
TYPE:	**FREQUENCY OF UPDATE:**
Data base	None

NUMBER AND TYPE OF RECORDS: 8,000 failure events and causes; Inventory data on ASME coded devices

DATA BOUNDARY: Safety and pollution prevention devices on offshore structures, e.g. subsurface safety valves

DATA ACCESS:

Contact: L.E. Bennett (Gene)
Minerals Management Services (OS-2)
U.S. Department of the Interior
P.O. Box 7944
Metairie, LA 70010
Phone: (504) 736-0557

DESCRIPTION:

The Failure and Inventory Reporting System (FIRS) program was developed by the Geological Survey Division of the U.S. Department of the Interior for safety and pollution prevention devices on offshore structures that produce or process hydrocarbons. The program collected data on mechanical and some electro-mechanical systems on offshore oil platforms. About 8,000 failure events were documented. Access has been limited to internal materials management system use. No real-time access or periodic output products have been available.

The system was composed of two programs. The safety device inventory reporting program provided information listing the number of safety and pollution devices by type, manufacturer, and model. The safety device failure reporting program provided information about the failures of these devices by failure causes, corrective actions, device type, manufacturer, model, and frequency of failure. Failure percentages, reliability and quality trends, mean time between repairs, and mean time between failures, along with other statistical information, were derived from these data.

The project was ended when financial support was discontinued; Mr. Bennett can field questions on the data set contents.

NON-PROCESS EQUIPMENT DATA BASES

TITLE:
Government Industry Data Exchange Program (GIDEP)

SPONSOR/AUTHOR: Various U.S. Government Agencies	**NO.:** 4.6-8
INDUSTRY: Varied	**TIME FRAME:** 1950's to Present
TYPE: Data base and Reports	**FREQUENCY OF UPDATE:** Continuous

NUMBER AND TYPE OF RECORDS: Recorded historical, experimental, and test data from 650 participating agencies/industries. May include failure frequencies, rates, modes, and mechanisms as well as repair times.

DATA BOUNDARY: Engineering, Reliability, Metrology, and Failure data on mechanical, electronic and electrical components.

DATA ACCESS:
Contact: Jim Richards or George Carver
GIDEP Operations Center
Corona, CA 91720
Phone: (714) 736-4677
Data accessibility: Standard outputs on microfilm and data base access (PC connection via modem) free to participants in GIDEP, who must agree to exchange (input) data; keyword searches by manufacturer for report abstracts

DESCRIPTION:
The Government-Industry Data Exchange Program (GIDEP) is funded and operated by the United States Government. GIDEP has approximately 650 participants from U.S. government agencies and various industries. All quality levels of components are represented with microcircuit, electronic, electrical, electromechanical, and mechanical systems included. The collected data specifically excludes both proprietary and confidential data; the emphasis is on off-the-shelf items. The data base is constructed from both recorded historical data furnished by the participants and from experimental or test data. Various environments, depending on the component's application, are included in the data bases; the operating environment is flagged. The failure frequencies, failure rates, failure modes, failure mechanisms, and mean times to repair may be included. GIDEP is comprised of four main data banks (referred to as data interchanges): (a) The engineering data interchange contains engineering evaluation and qualification test reports, nonstandard parts justification data, parts and material specifications, and other related engineering data on parts, components, materials, and manufacturing processes. (b) The reliability-maintainability data interchange contains failure rate/mode and replacement rate data on parts, components, and materials based on field performance information and/or reliability demonstration tests of equipment, subsystems, and systems. (c) The metrology data interchange contains metrology-related engineering data on test systems, calibration systems, and measurement technology and test equipment calibration procedures. (d) The failure experience data interchange contains objective failure information generated when significant problems are identified for parts, components, processes, fluids, materials, or safety and fire hazards.

Data Resource 4.6-8
Government Industry Data Exchange Program
Example Data Sheet

```
                                         FAILURE RATE DATA
                                         -----------------

                                                                                              90% GROUP CONFIDENCE INTERVAL
GENERIC CODE  NOMENCLATURE                                                                    LOWER       MEAN       UPPER
--------------------------                                                                    -----------------------------
                                                                                                                          N
REEL     PARTICI-                       OBSERVED                     STRESS      OBSERVATION                  TOTAL MILLION UNIT OF FAILURE  O
IMAGE    PANT     PART       MANUFAC-   ENVIRONMENT                  LEVEL   SCR PERIOD      MRT  NBR.   NER.  PART    MEA-   RATE     T
NUMBER   CODE     NUMBER     TURER      WITH MODIFIER  APPLICATION   ES-T-MS CLS BEGIN/END   MHRS FAIL   PARTS HOURS   SURE  /MILLION  E
------------------------------------------------------------------------------------------------------------------------------------------
928.55  VALVES, SEQUENCE                                                                      23.998   465.983  2210.624
---------------------------                                                                   ------------------------------
F003-0718 A6                            MSL-BOOST      LIQ RKT ENG   -M-H   A  6107/7406           1   580     0.002146 CYCS  465.983
    PREDOMINANT FAILURE MODES:  DEACTUATED PREMATURELY-2


928.60  VALVES, REGULATING OR REDUCING                                                        0.290   0.848   1.941
---------------------------                                                                   ------------------------------
F003-0223 S8 FS22475         GEM CO      GND-          RADAR         -M-       6803/7303           0     4   0.048556 HRS   <18.870
F003-0223 S8 211F84          HOKE INC    GND-          RADAR         -M-       6803/7303           0     3   0.120891 HRS    <7.579
F003-0223 S8 101C-A1         ROBERTSHAW  GND-          RADAR         -M-       6803/7303           0     2   0.024278 HRS   <37.741
F003-0223 S8 1G              DOLE VALVE  GND-          RADAR         -M-       6803/7303           0   110   1.882210 HRS    <0.486
F003-0223 S8 3047            DEMCO INC   GND-          RADAR         -M-       6803/7303           0     2   0.024278 HRS   <37.741
F003-0223 S8 R10GS           VAC ELEK    GND-          RADAR         -M-       6803/7303           0     3   0.127572 HRS    <7.182


928.60  VALVES, REGULATING OR REDUCING                                                        1.835   6.732   17.400
--------------------------------       HLI      REFER TO VOLUME ONE.                          ------------------------------


928.60  VALVES, REGULATING OR REDUCING                                                        80.866   96.111   113.488
---------------------------                                                                   ------------------------------
F003-1247 JF 9668M40         G.E.        JET-UNSLT     ACFT-ENGINE   - -M      7201/7509           3   361   0.386270 HRS     7.766
F003-1672 D4 39-170          HYDRO-AIRE  JET-MANND     BRAKE SYS     -M-       7401/7506          15   380   0.135280 HRS   110.881
    PREDOMINANT FAILURE MODES:  INTERNAL-9, LEAKING-4
F003-1672 D4 HP95610C-2      TELEDYNE    JET-MANND     WINGFOLD SYS  -M-       7401/7506           4   761   0.270916 HRS    14.764
    PREDOMINANT FAILURE MODES:  LEAKING-4


928.60  VALVES, REGULATING OR REDUCING                                                        17.462   19.410   21.523
--------------------------------       MBL      REFER TO VOLUME ONE.                          ------------------------------


928.60  VALVES, REGULATING OR REDUCING                                                        297.699   353.288   429.829
--------------------------------       SSJ      REFER TO VOLUME ONE.                          ------------------------------
```

NON-PROCESS EQUIPMENT DATA BASES

TITLE:
System Reliability Service

SPONSOR/AUTHOR:
UKAEA

NO.:
4.6-9

INDUSTRY:
Varied

TIME FRAME:
Unknown

TYPE:
Data Base and Reports

FREQUENCY OF UPDATE:
Continuous

NUMBER AND TYPE OF RECORDS: Status reports contain failure, repair, and maintenance data on 450 categories; data sets are broader

DATA BOUNDARY: A wide variety of equipment failure and performance records from utilities and manufacturers, plus collected inspection data

DATA ACCESS:

Contact: System Reliability Service, UKAEA
Culcheth, Warrington, Wigshaw Lane WA3 4NE
England
Phone: (044) 925-31244 Extension 520
Report accessibility: Limited to contributors

DESCRIPTION:

The Systems Reliability Service (SYREL) Data Base is used by the United Kingdom Atomic Energy Authority (UKAEA) Systems Reliability Service (SRS). It contains information from all industries, of which 20% is nuclear related. Information is submitted by utilities, licensing authorities, and manufacturers in addition to inspection data gathered by the SRS staff. The data base provides information to contributors on equipment performance, reliability, and availability, although certain aspects of manufacturing design are confidential. The data base is updated continuously and status reports are published every two months on 450 generic items.

Membership in SYREL entitles the user to the service of four other data banks:

1. The Accident Data Base

2. The Maloperation Data Base, covering events of less than accidents severity

3. The Event Data Store, applying specifically to nuclear power plants but discontinued because of high cost (approximately 1000 times normal rates)

4. The Piping Data Base, with a limited amount of detailed information which SYREL wishes to supplement with data from U.S. facilities

NON-PROCESS EQUIPMENT DATA BASES

TITLE:
SAIC Data Base

SPONSOR/AUTHOR:	**NO.:**
Joseph R. Fragola	4.6-10

INDUSTRY:	**TIME FRAME:**
Nuclear	1980 to Present
TYPE:	**FREQUENCY OF UPDATE:**
Data Bases and Reports	With new data availability

NUMBER AND TYPE OF RECORDS: Plant maintenance and repair records by system and component type; Selected basic event failure rates and unavailability

DATA BOUNDARY: Data mainly for pumps, valves, diesels, batteries, chargers, and heat exchangers

DATA ACCESS:

Contact: Joseph R. Fragola
SAIC, 342 Madison Ave., Suite 1100, NY, NY 10173
Phone: (212) 661-5780
Data Accessibility: Available for specific studies upon request
Cost: Based on hours and level of expertise required for study

DESCRIPTION:

In the course of performing reliability and risk analyses on nuclear plant systems and components, the New York office of SAIC has constructed numerous failure data sets. The primary (or preferred) source for the failure and repair information in all cases has been the plants' maintenance records for the entirety of their operating history (as available). To create a useful data set from these records, a series of sorts was performed that gradually structured the information by system, component group, component type, failure severity, failure mode, and failure cause. Hourly and demand-related failure rates and probabilities were then compiled for the study-specific components types and failure modes. The data is organized into folders documenting the development path from encoded records to calculational bases/assumptions to failure rates. Operational history for failure rate denominators is based upon plant shutdown/operating modes, system operation patterns, train/component interactions, testing schedules, and system diagrams.

The SAIC New York office houses five encoded data sets of the type described above and partial and raw data sets from five other plants. SAIC also maintains a published data source library from various industries. This library has been used to compile generic data sets for PRA screening, supplementing weak areas in plant-specific data sets, and as a basis for comparing plant data with world experience.

SAIC's data bank focuses on pumps, valves, and other common major components, but due to its depth has also been applied to studies on items such as strainers and bursting discs. Access is based on the scope of work and the estimated labor hours required to search and sort the data.

Data Resource 4.6-10
SAIC Data Base
Example Data Sheet

CARP -- DATA ANALYSIS SUMMARY REPORT

Type	Code	Component	Failure Mode	Unit	Dist	Mean	Lower	Median	Upper	P1	P2	Basis
AC	A	AIR COOLING UNIT	FAILS TO START	N	L	2.08-04	9.44-06	8.52-05	7.69-04	9.02+00		G
AC	F	AIR COOLING UNIT	FAILS TO RUN	H	L	9.05-06	1.09-06	5.55-06	2.82-05	5.08+00		G
AD	F	AIR DRYER	FAILS TO DELIVER FLOW	H	L	5.23-07	3.38-07	5.07-07	7.61-07	1.50+00		G
AF	F	AIR FILTER	FAILS TO DELIVER FLOW	H	L	7.23-06	1.16-06	4.91-06	2.08-05	4.24+00		G
AM	A	AIR COMPRESSOR	FAILS TO START	N	L	1.27-01	9.65-02	1.25-01	1.63-01	1.30+00		G
AM	F	AIR COMPRESSOR	FAILS TO RUN	H	L	2.48-03	2.74-05	5.09-04	9.49-03	1.86+01		G
AR	F	AIR RECEIVER	LOCAL FAULTS	H	L	6.00-07	1.04-08	1.56-07	2.32-06	1.49+01		G
AV	CN	AIR-OPERATED VALVE	FAILS TO OPERATE;OPEN OR SHUT	N	L	2.20-03	3.06-04	1.42-03	6.62-03	4.65+00		G
AV	KR	AIR-OPERATED VALVE	SPURIOUS OPERATION	H	L	3.02-06	1.46-07	1.27-06	1.11-05	8.70+00		G
B1	F	>4 KV BUS	FAULT	H	L	6.50-08	2.85-09	2.62-08	2.41-07	9.19+00		G
B2	F	<4 KV BUS	FAULT	H	L	1.32-07	3.59-09	4.27-08	5.07-07	1.19+01		G
B4	F	120 V BUS	FAULT	H	L	1.32-07	3.59-09	4.27-08	5.07-07	1.19+01		G
BC	F	BATTERY CHARGER	NO OUTPUT	H	L	7.60-06	2.84-07	2.84-06	2.85-05	1.00+01		G
BD	F	DC BUS	FAULT	H	L	6.50-08	2.85-09	2.62-08	2.41-07	9.19+00		G
BT	D	BATTERY	NO OUTPUT (DEMAND)	N	L	6.61-03	7.12-04	3.88-03	2.12-02	5.45+00		G

NON-PROCESS EQUIPMENT DATA BASES

TITLE:

The In-Plant Reliability Data Base for Nuclear Plant Components

SPONSOR/AUTHOR:

USNRC-RES

NO.:

4.6-11

INDUSTRY:

Nuclear

TIME FRAME:

1975 to 1984

TYPE:

Data Base and Reports

FREQUENCY OF UPDATE:

None

NUMBER AND TYPE OF RECORDS: Corrective maintenance records: 4000 for pumps; 5812 for valves; 698 for electrical failure rates and modes.

DATA BOUNDARY: Pumps, valves, diesels, battery charger, and inverters from 4 nuclear power plants.

DATA ACCESS:

Report ordering address: NTIS, Springfield, VA 22161
Phone: (703) 487-4650
Report cost: $19.95 each

DESCRIPTION:

The main objective of the In-Plant Reliability Data System (IPRDS) was to develop a comprehensive and component-specific data base for PRA and other component reliability-related statistical analysis. Data base personnel visited selected plants and copied all the plant maintenance work requests. They also gathered plant equipment lists and plant drawings and in some cases interviewed plant personnel for information on component populations and duty cycles. Subsequently, the maintenance records were screened to separate out the cases of corrective maintenance applying to particular components; these were reviewed to determine such things as failure modes, severity, and, if possible, failure cause. The data from these reports were encoded into a computerized data base.

Three reports have been issued containing IPRDS failure data. Information on pumps, valves, and major components in NPP electrical distribution systems has been encoded and analyzed. All three reports provide introductions to the IPRDS, explain failure data collections, discuss the type of failure data in the data base, and summarize the findings. They all contain comprehensive breakdowns of failure rates by failure modes with the results compared with WASH-1400 and the corresponding LER summaries. Statistical tables and plant-specific data are found in the appendixes. Because the data base was developed from only four nuclear power stations, caution should be used for other than generic application.

Allthe documents in this series have numbers beginning with NUREG/CR-, followed by these report-specific numbers: 2886 for pumps, 3154 for valves, 3831 for electrical, and 2641 for the methodology report.

Data Resource 4.6-11

In-Plant Reliability Data Base for Nuclear Plant Components

Example Data Sheet

Table A.22. In-plant reliability data system – service water booster pump

Failure rates by failure mode

Plant No. 1	Component class: pumps	Population: 4
Plant type: PWR	1st subclass: alternating	Average annual demands: 12
Operating period: 5 years	2nd subclass: centrifugal	Average duty cycle: 60 (C)
	3rd subclass: motor	Number of failures: 32
	4th subclass: service water booster	Population hours: 110,000
		Population demands: 240
		Maintenance frequency: 1.8E-4/h

Failure mode breakdown	Failure population	Time-related failure rate, λ_t				Demand-related failure rate, Q_d			
		Failures	Failures/10^6 h			Failures	Failures/10^3 cycles		
			Low	Recommended	High		Low	Recommended	High
Catastrophic	6								
Fails to start						1	0.21	4.2	20
Fails while running		5	18	45	95				
Degraded	6	6	24	55	110	6	11	25	49
Incipient	20	20	120	180	260	20	55	83	120

Failure cause breakdown (code)	Failure population	Failure population (%)	Repair time[a]			
			Population	Low	Mean	High
Seal/packing (3)	16	52	16	1.0	16.8	40.0
Filter/strainer plugged (21)	7	23				
Control circuit (37)	2	7	2	8.0	14.0	20.0
Unknown (0)	2	7				
Bearing (1)	1	3	1		1.5	
Motor (27)	1	3	1		20.0	
Switch (28)	1	3	1		4.0	
Leaky fitting (35)	1	3	1		2.0	

[a]Average time = 10 h; maximum time = 40 h; minimum time = 1.5 h.

NON-PROCESS EQUIPMENT DATA BASES

TITLE:
IEEE Standard 500-1984

SPONSOR/AUTHOR:
IEEE

NO.:
4.6-12

INDUSTRY:
Nuclear

TIME FRAME:
Varies

TYPE:
Data Base and Report

FREQUENCY OF UPDATE:
Every five years

NUMBER AND TYPE OF RECORDS: Over 100 data sheets, containing failure rate estimates for various failure modes and some repair times.

DATA BOUNDARY: Electrical, instrumentation, and mechanical components in nuclear power plants

DATA ACCESS:

Report ordering address: IEEE Service Center
445 Hoes Lane, Piscataway, NJ 08854

Report cost: IEEE Member—$121.50; Nonmember—$135.00

Report accessibility: No restrictions

DESCRIPTION:

The IEEE Standard 500 document was originally issued in 1977 as a manual of reliability data on electrical, electronic, and sensing components. Since that time, data have been gathered on mechanical equipment reliability, making the 1984 version of IEEE Std 500 a more comprehensive work by including data on over 1,000 components. The primary source for the reliability data presented in IEEE Std 500-1977 was the judgement of nearly 200 experts on component reliability in the nuclear power field. The Delphi method was used to compile and refine the expert opinion data. This information is included in IEEE Std 500-1984 for the nonmechanical components, but it is supplemented by data from 24 additional sources. These include nuclear industry data from published reports such as LER, NPRDS, and IPRDS data summaries and various manufacturers, utilities, and plants as well as component data from U.S. military sources. Therefore, the data developed represent either recorded data or the best collective judgment of a group of specialists. In each case, the source of the data is clearly identified. IEEE Std 500-1984 contains little unique data.

The IEEE Std 500 document is based on a hierarchical structure of component types set down in the manual's table of contents. The preface for each subsection (defined by a component type) provides a tree diagram that clearly shows the way the component classes have been subdivided to determine "data cells". The failure modes for each component class are also hierarchically organized according to failure severity: catastrophic, degraded, or incipient. Rates per hour and demand rates (per cycle) are both included, as well as upper and lower bounds.

Data Resource 4.6-12
IEEE Standard 500-1984
Example Data Sheet

(Composite of Ref 9, 10, 11 and 1, 10, 11, 24, 25)

CHAPTER: 11 Driven Equipment **SECTION:** 11.3 Generator Sets (Including Drivers) **SUBSECTION:** 11.3.1 Diesel

ITEM OR EQUIPMENT DESCRIPTION

FAILURE MODE	FAILURE RATE								(*) OUT OF SERVICE (†) REPAIR TIME OR (§) RESTORE (HOURS)			
	FAILURES/10^3 HOURS				FAILURES/10^3 CYCLES							
	LOW	REC	HIGH	REF	LOW	REC	HIGH	REF	LOW	REC	HIGH	REF
ALL MODES	0.10	1.18	50.0		0.30	18.2	528		0.25 †	9.76	480.0	
CATASTROPHIC	0.02	0.17	7.3		0.30	18.2	528					
Fails While Running	0.02	0.17	7.3									
Fails to Start					0.30	18.2	528					
DEGRADED	0.02	0.17	7.3									
Out of Spec	0.02	0.17	7.3									
INCIPIENT	0.06	0.84	35.4									

Reprinted from ANSI/IEEE Std. 500-1984.

NON-PROCESS EQUIPMENT DATA BASES

TITLE: Generic Data Base for Data and Models Chapter of the National Reliability Evaluation Program Guide

SPONSOR/AUTHOR:	**NO.:**
USNRC-RES	4.6-13

INDUSTRY:	**TIME FRAME:**
Nuclear	Unknown
TYPE:	**FREQUENCY OF UPDATE:**
Data Base and Report	None

NUMBER AND TYPE OF RECORDS: Estimates of hourly and per demand failure rates for 20 categories with error factors generated by experts

DATA BOUNDARY: Standard nuclear plant PRA equipment list

DATA ACCESS:

Contact: EG&G Idaho, Inc.
P.O. Box 1625, Idaho Falls, ID 83415
Report ordering address: Same as above
Report cost: Unknown

DESCRIPTION:

In April 1982, a data workshop was held to evaluate, discuss, and critique data in order to establish a consensus generic data set for the USNRC-RES National Reliability Evaluation Program (NREP). The data set contains component failure rates and probability estimates for loss of coolant accidents, transients, loss of offsite power events, and human errors that could be applied consistently across the nuclear power industry as screening values for initial identification of dominant accident sequences in PRAs. This data set was used in the development of guidance documents for the performance of PRAs.

TABLE 1. (continued)

Component and Failures Modes	Nominal Value	Error Factor	Remarks
4.5 Transformers			
4.5.1 All modes	1E-6/hr	3	
4.6 Emergency Diesel (Complete Plant)			Engine frame and associated moving parts, generator coupling, governor, static exciter, output breaker, lube oil system, fuel oil, intake and exhaust air, starting system; excludes starting air compressor and accumulator, fuel storage and transfer, load sequencers, and synchronizers. Failure to start is failure to start, accept load, and run for 1/2 hour; failure to run is failure to run for more than 1/2 hour, given start.
4.6.1 Failure to start	3E-2/d	3	
4.6.2 Failure to run, given start (emergency conditions)	1E-3/hr	10	
4.7 Relays			
4.7.1 Contacts fail to transfer (open or close)	1E-4/d	10	
4.7.2 Coil failure (open or short)	1E-6/hr	10	
4.8 Time Delay Relays			
4.8.1 Premature transfer	1E-4/d	10	
4.8.2 Fails to transfer			
4.8.2.1 Bimetallic	5E-6/hr	3	Non-consensus source. Data source is MIL-HDBK-217B. Fail-to-transfer rates are not currently available for non-bimetallic time delay relays.
4.8.2.2 Other	--	--	

NON-PROCESS EQUIPMENT DATA BASES

TITLE:

Offshore Reliability Data Handbook

SPONSOR/AUTHOR:	**NO.:**
OREDA Committee	4.6-14

INDUSTRY:	**TIME FRAME:**
Offshore Oil	1982 to 1984
TYPE:	**FREQUENCY OF UPDATE:**
Data Base and Report	None to date of this book

NUMBER AND TYPE OF RECORDS: Failure rates based on 14,000 failures for 300 component types. Mean values, upper and lower bounds are offered for different failure modes.

DATA BOUNDARY: Offshore oil equipment data and boundaries for safety systems, process systems, electrical systems, utility systems, cranes and drilling equipment.

DATA ACCESS:

Report ordering address: OREDA, P.O. Box 370, N-1322 HOVIK, Norway
Report cost: $150

DESCRIPTION:

The Offshore Reliability Data (OREDA) project began in 1982 as a Norweigan government effort to survey and evaluate existing data banks so that the best elements could be incorportated into a plan for offshore oil industry data collection. The operational phase of this project in 1983 was managed and funded by the OREDA Steering Committee, composed of oil company members. All the member companies contributed funds for the project which involved the monitoring, compilation, analysis, and presentation of data gathered from the individual companies. The results of the program were published in 1984 as the OREDA Handbook.

The OREDA Handbook is structured by the type of systems (safety, process, etc.) Components are cited beneath these headings and levels are designated by numeric subdivisions, such as 4.2.1 and 4.2.2, much like IEEE Std. 500-1984. A preface section precedes each system and each component type data page set. This system parameters to be considered are listed. A full structured taxonomy, including components and subclasses numerically identified, is also provided. Each component level preface describes a variety of parameters including operating mode, component specifications, and a diagram showing the component boundary. The data table pages all employ the same format in which the heading provides component information (operating hours, component population, etc.) and the body presents the failure and repair data. Mean values for failure data and upper and lower bounds are correlated to failure modes structured much like those in IEEE Std. 500-1984, except that the "Catastrophic" severity category is called "Critical" by OREDA. Repair time is given for each mode as well.

Data Resource 4.6-14

Offshore Reliability Data (OREDA-84)

Example Data Sheet

Reprinted from OREDA-84, with permission.

Taxonomy no. and item
1.3.1.1
Safety Systems
Fire-Fighting Systems
Fire Water Pumps
Diesel Motor, Gear/Shaft Driven Pumps

Description
Multistage centrifugal pumps driven by diesel engine. They are driven through a right angle gear box on approximately half of the population, and through a straight vertical gear box on the rest. Battery/motor start with air or hydraulic motor back-up. Automatic start from pressure drop in fire main with alternative manual facilitiy. Typical output 450 m^3/h and 700—1500 m^3/h, 120—150 m head.

Application
Feeds the fire-water ring main on platform with water from the sea.

Operational mode
Standby, intermittent operation.

Internal environment
Sea-water.

External environment
Engine, gearbox and controls installed in ventilated and heated modules. Steel tube protects pumps and drive shaft. Saline atmosphere.

Testing and maintenance
Weekly: Check coolant and oil levels.
Test run and check for leaks.
3 months: Change oil and filter elements, clean magnetic plugs, check engine manifold and water pump for leaks, lubricate, clean crankshaft vent valve.
6 months: Increasingly stringent maintenance.

Failure causes
Start failures predominantly on instrumentation. Running failures on seals, bearings, drive components and instruments.

Item boundary specification
Only failures within the boundary indicated with the dash line in the figure below are included in the reliability data source.

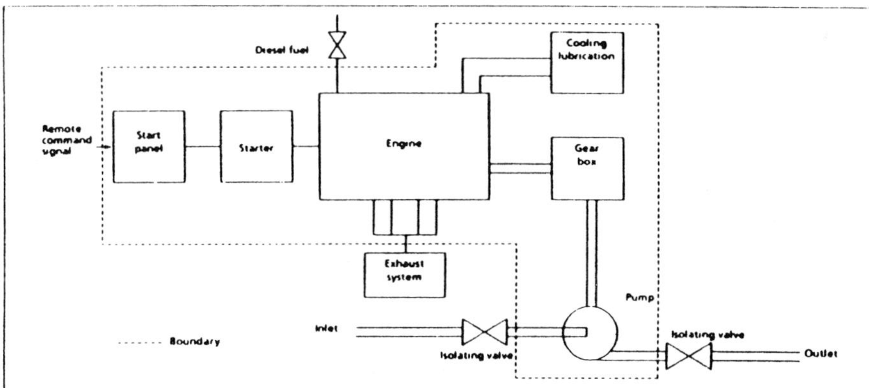

Fig 4.30

Taxonomy no	**Item**
1.3.1.1	Safety Systems Fire-Fighting Systems Fire Water Pumps Diesel Motor, Gear/Shaft Driven Pumps

Population	**Samples**	**Aggregated time in service (10^6 hours)**		**No of demands**
		Calendar time *	**Operational time** †	
17	10	.3826	.0002	1135

Failure mode	**No of failures**	**Failure rate (per 10^6 hrs)**			**Active repair (hours)**	**Repair (manhours)**		
		Lower	**Mean**	**Upper**		**Min**	**Mean**	**Max**
Critical	**80 * 13 †**	**120 26000**	**210 47000**	**310 78000**	**-**	**-**	**86**	**-**
Failed to start	75 * 9 †	100 6200	190 32000	290 69000	-	24	86	120
Failed while running	5 * 4 †	2.0 4600	23 15000	51 36000	-	3	93	130
Degraded	**24 * 3 †**	**30 0**	**71 14000**	**120 45000**	**-**	**-**	**180**	**-**
High temperature	22 * 3 †	22 0	66 14000	120 44000	-	6	190	400
Low output	1 *	0.14	2.6	12	-	-	-	-
Unknown	1 *	0.14	2.6	12	-	-	96	-
Incipient	**30 * 20 †**	**56 41000**	**150 81000**	**250 130000**	**-**	**-**	**15**	**-**
Vibration	1 * 1 †	0.0092 0	3.6 3800	15 19000	-	-	46	-
Leakage	20 * 10 †	13 7100	71 35000	140 73000	-	11	15	17
Unknown	9 * 9 †	0.20 0	77 42000	160 110000	-	1	11	12
Unknown	**169 * 9 †**	**270 0**	**400 37000**	**550 86000**	**-**	**-**	**72**	**-**
Failed	169 * 9 †	270 0	400 37000	550 86000	-	3	72	95
All modes	**303 * 45 †**	**680 87000**	**840 180000**	**1000 280000**	**-**	**-**	**81**	**-**

Comments

On demand probability (per 1000 demands). Lower : Mean : Upper

Critical - Failed to start: 0.37 : 50 : 130

Note: The estimates were in some cases based on different subsets of samples for the different failure modes. This results from tests of statistical consistence among samples, see sect 3.4.

NON-PROCESS EQUIPMENT DATA BASES

TITLE:
RADC Non-Electronic Reliability Notebook

SPONSOR/AUTHOR:
Rome Air Development Center

NO.:
4.6-15

INDUSTRY:
Military

TIME FRAME:
July 1982 to August 1985

TYPE:
Data Base and Report

FREQUENCY OF UPDATE:
None

NUMBER AND TYPE OF RECORDS: Failure rates based on operations and maintenance records and industry-wide sources

DATA BOUNDARY: Covers 250 non-electronic part types such as actuators, complings, filters for 12 classes of environment (air, ground, sea) used in military systems and equipment.

DATA ACCESS:

Contact: James A Collins, RADC Griffiss AFB, NY 13441-5700
Phone: (315) 330-4810 or 4726
Report order address: National Technical Information Service (NTIS)
Springfield, VA 22161
Phone: (703) 487-4650
Report No.: AD/A163 900
Report cost: $50.95

DESCRIPTION:

The objective of this notebook was the collection, analysis, and presentation of nonelectronic component failure data and the presentation of analytical methods that form the state-of-the-art in nonelectronic reliability analysis. This report replaces the former Nonelectronic Reliability Notebook (RADC-TR-75-22).

This notebook is divided into six sections: Introduction, Applicable Statistical Methods for Nonelectronic Reliability, Reliability Specifications, Special Application Methods for Reliability Prediction; Part Failure Characteristics, and Reliability Demonstration Tests. Data is located in section 5.0 on Part Failure Characteristics. This section describes the results of the statistical analyses of failure data from more than 250 distinct nonelectronic parts collected from recent commercial and military projects. This data was collected in-house (from operations and maintenance reports) and from industry wide sources. Tables, alphabetized by part class/ part type, are presented for easy reference to part failure rates assuminng that the part lives are exponentially distributed (as in previous editions of this notebook, the majority of data available included total operating time, and total number of failures only). For parts for which the actual life times for each part under test were included in the database, further tables are presented which describe the results of testing the fit of the exponential and Weibull distributions.

Data Resource 4.6-15

RADC Non-Electronic Reliability Notebook

Example Data Sheet

BATTERY

COMPONENT PART TYPE	APPL ENV	USER CODE	FAILURE RATE PER MILLION HRS. POINT ESTIMATE	60% UPPER SINGLE-SIDED	20% LOWER INTERVAL	80% UPPER INTERVAL	# OF RECS	# OF FAIL	OPERATING HRS (E6)
RECHARGEABLE	A	C	-----	0.159	-----	-----	1	0	5.760
RECHARGEABLE	A	M	348.852	-----	342.921	355.955	3	2810	8.055
RECHARGEABLE	AUT	C	248.851	-----	238.745	259.459	1	433	1.740
RECHARGEABLE	DOR	C	0.006	-----	0.003	0.014	1	2	313.175
RECHARGEABLE	DOR	M	0.016	-----	0.013	0.022	3	12	732.564
RECHARGEABLE	GF	?	1.630	-----	1.302	2.044	3	18	11.042
RECHARGEABLE	GF	M	1.498	-----	1.046	2.138	1	8	5.339
RECHARGEABLE	GM	M	15.748	-----	6.406	34.040	2	2	0.127
RECHARGEABLE	HEL	M	676.768	-----	636.364	720.148	2	201	0.297

BEARING

COMPONENT PART TYPE	APPL ENV	USER CODE	FAILURE RATE PER MILLION HRS. POINT ESTIMATE	60% UPPER SINGLE-SIDED	20% LOWER INTERVAL	80% UPPER INTERVAL	# OF RECS	# OF FAIL	OPERATING HRS (E6)
BALL	A	C	1.372	-----	0.272	4.171	1	1	0.729
BALL	A	M	5.133	-----	4.787	5.507	2	158	30.784
BALL	AI	M	4.829	-----	3.799	6.148	1	16	3.313
BALL	AIF	M	19.323	-----	9.863	35.795	3	3	0.155
BALL	AIT	M	-----	0.723	-----	-----	1	0	1.267
BALL	DOR	M	0.010	-----	0.007	0.014	3	9	903.040
BALL	GF	C	13.985	-----	10.003	19.511	1	9	0.643
BALL	GF	M	1.148	-----	1.001	1.319	8	44	38.320
BALL	GM	M	0.094	-----	0.054	0.159	1	4	42.554
BALL	HEL	M	13.398	-----	10.963	16.408	2	22	1.642
BALL	NS	M	-----	0.053	-----	-----	1	0	17.134
BALL	NSB	M	4.728	-----	1.923	10.220	1	2	0.423
BALL	SF	M	-----	0.688	-----	-----	2	0	1.332

NON-PROCESS EQUIPMENT DATA BASES

TITLE:
Reliability Prediction of Electronic Equipment (Military Handbook 217E)

SPONSOR/AUTHOR:
RADC

NO.:
4.6-16

INDUSTRY:
Government and Military

TIME FRAME:

TYPE:
Data Base and Report

FREQUENCY OF UPDATE:
Every 12 to 18 months

NUMBER AND TYPE OF RECORDS: Failure rates for 13 categories of components.

DATA BOUNDARY: Electronic components used in military electronic systems and equipment.

DATA ACCESS:

Contact: Anthony Fedducia
Rome Air Development Center
Attn: RBET
Griffiss Air Force Base, NY 13441
Phone: (315) 330-4921
Report cost: None
Report accessibility: Unrestricted

DESCRIPTION:

Military Handbook 217E (MIL 217E) establishes uniform methods for predicting the reliability of military electronic equipment and systems. There are two methods of reliability predictions, namely parts count and parts stress analysis.

The part stress analysis prediction section contains failure rate models for a broad variety of parts used in electronic equipment. This method includes the effects of part quality factors and environmental factors. The tabulated values of the base failure rate are "cut off" at the design temperature and stress of the part.

The parts count method is suitable for early design phase reliability prediction. The method uses information on generic types, quality levels, and environment. The latter two effects are considered with the application of specified factors. The failure rates for both methods are calculated using the same generic expressions.

The military electronic components are subdivided into 13 categories. Generic failure rates for these categories are derived by applying single-value factors, such as part quality level and operating environment. The tabulated base failure rates are terminated at the rated design temperature and material stress of the part. The handbook provides example failure rate calculations for all of the categories considered as well as for miscellaneous related parts. Maintenance information is covered by a separate program, which uses data from MIL 470 for equipment testing and MIL 471 for prediction of maintenance.

CAPACITORS
MIL-C-55514, CFR

SPECIFICATION	STYLE	DESCRIPTION
MIL-C-55514	CFR	Plastic, Metallized Plastic, ER

Part operating failure rate model (λ_p):

$$\lambda_p = \lambda_b \; (\pi_E \times \pi_Q \times \pi_{CV}) \text{ failures}/10^6 \text{ hrs.}$$

where the factors are shown in Tables 5.1.7.1-28 through -33.

TABLE 5.1.7.1-28
Environmental Mode Factors

ENVIRON-MENT	π_E
G_B	1
G_{MS}	1.1
G_F	1.9
G_M	9.3
M_P	11
N_{SB}	5
N_S	5.7
N_U	16
N_H	16
N_{UU}	17
A_{RW}	23
A_{IC}	3.5
A_{IT}	6.5
A_{IB}	9.5
A_{IA}	6.5
A_{IF}	15
A_{UC}	10
A_{UT}	20
A_{UB}	25
A_{UA}	20
A_{UF}	40

ENVIRON-MENT	π_E
S_F	1
M_{FF}	11
M_{FA}	15
U_{SL}	31
M_L	36
C_L	610

Table 5.1.7.1-29
Base Failure Rate Tables
for Capacitor Spec and Style

Spec MIL-C	Style	λ_b Table Number
55514	Char. M, N	5.1.7.1-32
	Char. Q, R, S	5.1.7.1-33

4.7 INDEX OF NON-PROCESS EQUIPMENT DATA SOURCES

4.7 INDEX OF NON-PROCESS EQUIPMENT DATA SOURCES

NO.	TITLE	INDUSTRY	NO. & TYPE OF RECORDS	DATA BOUNDARY	PAGE
4.7-12	A Statistical Analysis of Nuclear Power Plant (Pump and Valve) Failure Rate Variability: Some Preliminary Results	Nuclear	All IPRDS data base records for the pumps and valves selected for analysis	The set of valves and pumps selected for analysis from the IPRDS data base	104.
4.7-13	Investigation of Valve Failure Problems in LWR Power Plants	Nuclear	195 LERs valve failures causing trips from 12/72 to 12/78, plus all valve failures for 10 stations from 2/66 to 1/79	Valve-related events reported in LERs, as noted above	105.
4.7-14	The Reliability of Emergency Diesel Generators at U.S. Nuclear Power Plants	Nuclear	Number of failures and demands for 154 diesels	Failure to start and failure to load and run data for Diesel Generators	106.
4.7-15	EPRI Guide	Power	3000 report descriptions	Advanced power, coal, electrical, nuclear, energy management, and environment topic areas	107.
4.7-16	Component Failure and Repair Data for Gasification-Combined-Cycle Power Generation Units	Power	Failure rates and averages restore times from published, analytical, and judgment data	Data for 121 system/component groups from Coal Gasification Combined-Cycle Units	108.
4.7-17	Performance of Pipework in the British Sector of the North Sea	Offshore Oil and Natural Gas	Failure rates based on 27 actual incidents from UK DOE reports	Offshore oil, gas, and process fluid submarine pipelines within the UK Continental Shelf	109.
4.7-18	Reliability Analysis Center Handbooks	Government and Military	Data summaries of hundreds of records by component and environment	Electronic component reliability data, i.e. microelectronic devices, high technology components	110.
4.7-19	An Analysis of Reportable Incidents for Natural Gas Transmission and Gathering Lines 1970 through June 1984	Natural Gas	Several thousand incidents of service and test failures	Gas pipelines	111.
4.7-20	Severities of Transportation Accidents	Automotive, Airline, Truck	Graphs and tables giving probability of accidents having certain severities	US transportation industry	112.
4.7-21	Pressure Vessel Failure Statistics and Probabilities	Nuclear	4 tables containing failure data for vessels	Primarily concerned with boiler failures	113.
4.7-22	Characteristics of Pipe System Failures in Light Water Reactors	Nuclear	Approximately 100 records of pipe failure rates in a wide variety of failure modes	Nuclear Power Plant Piping	114.
4.7-23	Reliability of Emergency AC Power Systems	Nuclear	900 occurrences of diesel generator failures at U.S. nuclear power plants	Diesel generator performance data for 18 different nuclear power plants	115.

NON-PROCESS EQUIPMENT DATA SOURCES

TITLE:

An Aging Failure Survey of LWR Safety Systems and Components

SPONSOR/AUTHOR: EG&G Idaho

NO.: 4.7-1

INDUSTRY: Nuclear

TIME FRAME: Approx. 1970 - 1986

TYPE: Report

FREQUENCY OF UPDATE: None

NUMBER AND TYPE OF RECORDS: 4 Tables of component failures per years of service

DATA BOUNDARY: Light Water Reactor Safety System Components

DATA ACCESS:

Contact: Babette Meale
EG&G Idaho, INEL, P.O. Box 1625, Idaho Falls, ID 83145
Phone: (208) 526-9978
Ordering Address: NTIS, Springfield VA 22161
Phone: (703) 487-4650
Report No.: NUREG/CR-4747 Vols. 1 and 2, 7/87 and 6/88
Report Cost: Publication imminent, no cost info. to date

DESCRIPTION:

This report describes the methods, analyses, results, and conclusions of two different aging studies. The first study consists of a survey of light water reactor component failures associated with 15 selected safety and support systems. Analysts used computerized sorting techniques to classify component failures into generic failure categories. The second study consists of careful examination of component failure records to identify and categorize the reported cause of component failures. The systems evaluated in the failure-cause analysis were the auxiliary feedwater, Class 1E electrical power distribution, high-pressure injection, and service water. Tables and figures are presented, indicating the systems and the components within those systems most affected by aging. Also provided are engineering insights drawn from the data. The Volume 2 report presents all of the Volume 1 data from FY-86 combined with the data gathered in FY-87. Data was taken from the NPRDS system from the failure categories of human error, design-related and aging-related. The number of failures of each component type per category per service time is given in the appendices.

NON-PROCESS EQUIPMENT DATA SOURCES

TITLE: Analysis of Dependent Failure Events and Failure Events Caused by Harsh Environment Conditions

SPONSOR/AUTHOR: USNRC-RES	**NO.:** 4.7-2
INDUSTRY: Nuclear	**TIME FRAME:** Intermittent: January 1971 to December 1985
TYPE: Report	**FREQUENCY OF UPDATE:** None

NUMBER AND TYPE OF RECORDS: 700 events representing common cause failures and failures caused by harsh environments

DATA BOUNDARY: Licensee Event Reports on failures of 26 component and subcomponent types listed below

DATA ACCESS:

Contact: Michael Bohn, Sandia National Laboratories (SNL), Division 6412
P.O. Box 5800, Albuquerque, NM 87185-5800
Phone: (505) 966-5232 (FTS 844-5232)
Report ordering address: Same as above
Report No.: JBFA-LR-111-85, 8/85 Report cost: Unknown
Report accessibility: This report is the property of SNL and must be obtained through SNL

DESCRIPTION:

This is a letter report from JBF Associates Inc., to Sandia National Laboratories (SNL) summarizing JBF's efforts to analyze dependent (common cause) failures and failures caused by harsh environments. The information used for the analysis was taken from over 1000 failure reports (mostly abstracts of LERs that were assembled for other studies). The 26 groups of components selected for study are: accumulators, batteries, cables, control rod drives, dampers, diesel generators, drains, air filters, fuel bundles, heat exchangers, heaters, instrumentation and controls (I&C), motors, offsite power supplies, access penetrations, piping, condensate polishers, pumps, electrical equipment, scram discharge volumes, shock suppressors, spargers and nozzles, strainers, transformers, valves, and other. Air filters include moisture separators, steam traps, high efficiency particulate air (HEPA) filters, charcoal filters, roughing filters, and prefilters. The I&C category includes sensors, switches, detectors, monitors, transmitters, cables, amplifers, bistables, calculators, camparators, and summators. Electrical equipment includes relays, breakers, starters, and timers. The "other" category applies to those events in which the components involved are not explicitly indicated. 700 unique events of interest were identified and sorted into 10 data sets for analysis. The data sets were constructed with different areas of focus for dependent failure (e.g., PWR-BWR comparisons, component type harsh environment). The description for each event includes LER number, NSIC number, date, plant, event description, the failure cause, common cause classification, component type, and, in several cases, the number of failed/unfailed components.

NON-PROCESS EQUIPMENT DATA SOURCES

TITLE:
Emergency Diesel Generator Operating Experience

SPONSOR/AUTHOR:
USNRC-NRR

NO.:
4.7-3

INDUSTRY:
Nuclear

TIME FRAME:
January 1981 to December 1983

TYPE:
Report

FREQUENCY OF UPDATE:
None

NUMBER AND TYPE OF RECORDS: 600 occurrences of failure from event reports and questionnaires

DATA BOUNDARY: Diesel generators

DATA ACCESS:
Contact: Ronald E. Battle, Oak Ridge National Laboratory
Building 3500, MS-8, Oak Ridge, TN 37831
Phone: (615) 574-5531 (FTS 624-5531)
Report ordering address: NTIS, Springfield, VA 22161
Phone: (703) 487-4650
Report No.: NUREG/CR-4347 Report cost: $14.95

DESCRIPTION:

The purpose of this report is to update the analysis of the operating experience of emergency DGs in nuclear power plants contained in NUREG/CR-2989.

The LER data base served as the primary source of DG failure data, while a data base for DG successes was formed from nuclear plant licensees' responses to a USNRC questionnaire (Generic Letter 84-15). Estimates of DG failure on demand were calculated from the LER data, DG test data, and response data from the questionnaire. The questionnaire also provided data on DG performance during complete and partial LOSP and in response to safety injection actuation signals. Trends in DG performance are profiled. The effects of testing schedules on diesel reliability are assessed. Individual failures are identified in an appendix.

NON-PROCESS EQUIPMENT DATA SOURCES

TITLE: A Review of Issues to Improving Nuclear Power Plant Diesel Generator Reliability

SPONSOR/AUTHOR: USNRC-NRR

NO.: 4.7-4

INDUSTRY: Nuclear

TIME FRAME: January 1980 to December 1985

TYPE: Report

FREQUENCY OF UPDATE: None

NUMBER AND TYPE OF RECORDS: 5000 events from responses to USNRC and Brookhaven National Laboratory (BNL) questionnaires

DATA BOUNDARY: Diesel generators

DATA ACCESS:

Contact: James Higgins, Brookhaven National Laboratory, Building 130
Upton, NY 11973
Phone: (516) 282-2432 (FTS 666-2432)
Report ordering address: NTIS, Springfield, VA 22161
Phone: (703) 487-4650
Report No.: NUREG/CR-4557, 4/86 Report cost: $25.95

DESCRIPTION:

The report provides an analysis of data received from utilities in reponse to the USNRC Generic Letter 84-15. Inputs obtained through responses to a BNL questionnaire designed specifically for the study were also included in the analysis. Recommendations made for DG reliability by other groups including industry organizations (such as INPO and ASME), DG manufacturers or vendors, foreign DG users, and the Advisory Committee on Reactor Safeguards (ACRS) were also evaluated.

Report recommendations include ways to improve DG reliability and maintenance programs. Other information provided includes: the DG reliability at every site for the last 20 starts and last 100 starts, a listing of the number of DGs per unit, summaries of responses to generic letter 84-15 including population data, and summaries of individual utility reponses to the BNL questionnaire.

NON-PROCESS EQUIPMENT DATA SOURCES

TITLE: Evaluation of Diesel Unavailability and Risk Effective Surveillance Test Intervals

SPONSOR/AUTHOR: USNRC-NRR

NO.: 4.7-5

INDUSTRY: Nuclear

TIME FRAME: 1981 to 1987

TYPE: Report

FREQUENCY OF UPDATE: None

NUMBER AND TYPE OF RECORDS: DG test and accident unavailability for 10 operating years

DATA BOUNDARY: Diesel generators

DATA ACCESS:

Contact: Pranab Samanta, Brookhaven National Laboratory, Building 130
Upton, NY 11973
Phone: (516) 282-2123 (FTS 666-2123)
Report ordering address: NTIS, Springfield VA 22161
Phone: (703) 487-4650
Report No.: NUREG/CR-4810, 5/87 Report Cost: $14.95

DESCRIPTION:

This report presents an analysis of DG unavailability, caused both by failure occurring while the DG is on standby and test-caused failures. The report presents a methodology for determining testing intervals (TIs) so that diesel unavailability is at an acceptably low level. Sensitivity analyses of test unavailability to varying TIs are presented.

PC-based models are presented for evaluating diesel unavailability. Parameters for the models are discussed in the report, but individual DG unavailability events are not listed. Generic TIs for a range of parameters and population of plants are displayed.

The software for the PC-based algorithm evaluating the effect of varying TI on unavailability is transferable by PC disk. The software required to support the analysis is Lotus 1-2-3. One hypothetical set of parameters for the model is included on the disk.

NON-PROCESS EQUIPMENT DATA SOURCES

TITLE:

Operating Experience and Aging--Seismic Assessment of Electric Motors

SPONSOR/AUTHOR:	**NO.:**
USNRC-RES	4.7-6

INDUSTRY:	**TIME FRAME:**
Nuclear	January 1974 to December 1983

TYPE:	**FREQUENCY OF UPDATE:**
Report	None

NUMBER AND TYPE OF RECORDS: Over 500 events representing occurrences of electric motor failure in nuclear power plants

DATA BOUNDARY: Failures of electric motors

DATA ACCESS:

Contact: Manomohan Subudhi, Brookhaven National Laboratory, Building 130
Upton, NY 11973
Phone: (516) 282-2429 (FTS 666-2429)
Report ordering address: NTIS, Springfield, VA 22161
Phone: (703) 487-4650
Report No.: NUREG/CR-4156, 6/86 Report cost: Unknown

DESCRIPTION:

This report provides an aging assessment of electric motors and was conducted under the auspices of the USNRC NPAR. Pertinent failure-related information was derived from LERs, IPRDS, NPRDS, and NPE including failure modes, mechanisms, and causes for motor problems. In addition, motor design and materials of construction were reviewed to identify age-sensitive components. The study included consideration of the seismic susceptibility of age-degraded motor components to externally-induced vibrational effects.

The aforementioned reviews and assessments were assimilated to characterize the effect of dielectric, rotational, and mechanical hazards on motor performance and operational readiness. Functional indicators were identified that can be monitored to assess motor component deterioration caused by aging or other accidental stressors. The study also includes a preliminary discussion of current standards and guides, maintenance programs, and research activities pertaining to nuclear power plant safety-related electric motors. Included are motor manufacturer recommendations, responses from repair facilities to a questionnaire, in-service inspection data, expert knowledge, USNRC-IE audit reports, and standards and guides published by the Institute of Electrical and Electronics Engineers (IEEE).

NON-PROCESS EQUIPMENT DATA SOURCES

TITLE:

A Review of Emergency Diesel Generator Performance at Nuclear Power Plants

SPONSOR/AUTHOR:

USNRC-IE

NO.:

4.7-7

INDUSTRY:

Nuclear

TIME FRAME:

January 1980 to December 1983

TYPE:

Report

FREQUENCY OF UPDATE:

None

NUMBER AND TYPE OF RECORDS: 500 occurrences of DG failure reported in LERs, 10 CFR 50.55E, Part 21, NPRDS, and EPRI document files

DATA BOUNDARY: Diesel generators

DATA ACCESS:

Contact: James Higgins, Brookhaven National Laboratory, Building 130
Upton, NY 11973
Phone: (516) 282-2432 (FTS 666-2432)
Report ordering address: NTIS, Springfield, VA 22161
Phone: (703) 487-4650
Report No.: NUREG/CR-4440, 11/85 Report cost: $12.95

DESCRIPTION:

This report evaluates recent performance of DGs and all DG vendors with the exception of Transamerica Delaval, Inc. (TDI), because of the emphasis already being given to TDI diesels in other studies. For the period 1980 through 1983 inclusive, BNL reviewed and evaluated DG failure data, DG vendor inspection reports, the TDI lessons learned as they related to the other vendors, and previous pertinent studies. The data sources used for DG failure analysis were LERs, 10 CFR 50.55E, Part 21, NPRDS, and EPRI document files. The DG failures were classified relative to the DG component that failed (e.g., main bearings, starting system). The failures were also categorized and analyzed by mode, manufacturer, and cause. Manufacturers with significant failures are identified in the report.

NON-PROCESS EQUIPMENT DATA SOURCES

TITLE: Data Summaries of Licensee Event Reports at U.S. Commercial Nuclear Power Plants (Various Components)

SPONSOR/AUTHOR:	**NO.:**
USNRC-RES	4.7-8

INDUSTRY:	**TIME FRAME:**
Nuclear	January 1972 to December 1983

TYPE:	**FREQUENCY OF UPDATE:**
Report	None

NUMBER AND TYPE OF RECORDS: 11209 one-line event descriptions on specific component types; failure rates and error factors

DATA BOUNDARY: Pumps,valves, diesels inverters, relays, circuit breakers (in separate reports)

DATA ACCESS:

Contact: Mr. Mike Trojovsky, EG&G Idaho, Inc., P.O. Box 1625
Idaho Falls, ID 83415
Phone: (208) 526-9098 (FTS 583-9098)
Report ordering address: NTIS, Springfield, VA 22161
Phone: (703) 487-4650
Report cost: Range from $20 to $45

DESCRIPTION:

EG&G Idaho's Idaho National Engineering Laboratory reviewed Licensee Event Reports (LERs), both qualitatively and quantitatively, to extract reliability information in support of the USNRC's effort to gather and analyze component failure data for U.S. commercial nuclear power plants. LERs describing failures or command faults (failure due to lack of needed input) for selected components have been analyzed in this program. Separate reports have been issued for batteries and battery chargers, control rods and drive mechanisms, diesel generators, I&C, Inverters, primary containment penetrations, protective relays and circuit breakers, pumps, and valves.

The body of each report has two major parts: the methodology used in encoding the LERs and a summary of results also containing fault and failure rates for significant component-fault mode combinations. Denominator information for the rates comes from plant final safety analysis reports (FSARs) together with very coarse estimates of numbers of demands. Specific plant fault data were averaged to obtain rates for the four NSSS vendors, for PWRs, for BWRs, and for the aggregate of both reactor types. Chi-squared bounds for the rates are computed. Appendices in the reports provide explanations of LER reporting variations, the LER coding scheme, and the methods used to estimate the fault rates.

All the document numbers begin with NUREG/CR-, followed by these report-specific numbers: 1362 for diesels, 3867 for inverters, 1205 for pumps, 1740 for I&C, and 1363 for valves.

NON-PROCESS EQUIPMENT DATA SOURCES

TITLE:

Pipe Break Frequency Estimation for Nuclear Power Plants

SPONSOR/AUTHOR:	**NO.:**
USNRC-RES	4.7-9

INDUSTRY:	**TIME FRAME:**
Nuclear	Through 1984
TYPE:	**FREQUENCY OF UPDATE:**
Report	None

NUMBER AND TYPE OF RECORDS: 19 occurrences of pipe failures (breaks), supplemented by expert-opinion estimates.

DATA BOUNDARY: Leaks of 1 gpm for 2 inches in diameter pipe; 50 gpm for all pipe for 81 nuclear plants.

DATA ACCESS:

Contact: Ronald E. Wright, EG&G Idaho, Inc., P.O. Box 1625
Idaho Falls, ID 83415
Phone: (208) 526-9467 (FTS 583-9467)
Report ordering address: NTIS, Springfield, VA 22161
Report No.: NUREG/CR-4407, 5/87
Report cost: $24.00

DESCRIPTION:

The study empirically develops frequencies and bounds for safety-significant pipe failures in commercial NPPs. Its purpose is to update the pipe break frequencies reported in the Reactor Safety Study (WASH-1400), which are used in many risk analyses. The study involved the review of various data sources for actual piping failure events of significant magnitude. The data sources reviewed were LERs, NPE, and several other sources documented in the report. Information was extracted concerning conditional factors such as the system in which the failure occurred, the operational mode of the plant, and the size of the pipe involved to permit estimation of conditional pipe break frequencies useful to risk analysts. Because there have been few significant pipe failures, the sparse real data was supplemented with expert-opinion data. The report presents the results of combining the real and subjective data through Bayesian statistical methods. That is, posterior probabilities of given failure rates were determined and are presented in the report. The rates of pipe failures are also analyzed to determine whether or not the rates are dependent on the system under consideration, the operational mode of the plant, the size of the pipe, or other factors. In statistical terms, an analysis of variance assessment was made on the rates of pipe failure.

NON-PROCESS EQUIPMENT DATA SOURCES

TITLE:
ATWS: A Reappraisal,Part 3: Frequency of Unanticipated Transients

SPONSOR/AUTHOR:
USNRC-NRR

NO.: 4.7-10

INDUSTRY:
Nuclear

TIME FRAME:
January 1969 to December 1984

TYPE:
Report

FREQUENCY OF UPDATE:
None

NUMBER AND TYPE OF RECORDS: 200 pump failure events from Arkansas Nuclear Unit 1, Calvert Cliffs Unit 1, and Indian Point Unit 3 nuclear plants

DATA BOUNDARY: Nuclear reactor coolant pump seals

DATA ACCESS:

Contact: M. Ali Azarm, Brookhaven National Laboratory
Department of Nuclear Energy, Upton, NY 11973
Phone: (516) 282-4922 (FTS 666-4922)
Report ordering address: NTIS, Springfield, VA 22161
Phone: (703) 487-4650
Report No.: NURÉG/CR-4400, 12/85 Report cost: $19.95

DESCRIPTION:

This report briefly describes a group of reactor coolant pump (RCP) seal failures that occurred at Arkansas Nuclear Unit 1, Calvert Cliffs Unit 1, and Indian Point Unit 3. Both mechanical and maintenance-induced RCP failure are discussed. For each event, the following information is provided: the date, the pump identification code, the nature of the failure, the maximum leakage per minute, and the total leakage in gallons. Data sources used as input were LERs, NPRDS, and final safety analysis reports (FSARs).

The report includes pedigree information on each plant. This includes plant name and unit number, type, vendor, number of pumps, pump designer, pump model number, and number of seal stages.

The report presents the findings from the analysis of the RCP failures. Estimates of the annual frequency for the spectrum of leak rates induced by RCP seal failures and their impact on plant safety (contribution to core-melt frequency) are made. The safety impact of smaller RCP seal leaks was assessed qualitatively, whereas for leaks above the normal makeup capacity, formal PRA methodologies were applied. Also included are the life distribution of RCP seals and the conditional leak rate distributions, given a RCP seal failure; the contribution of various root causes; and estimates for the dependency factors and the failure intensity for the different combinations of pump designers and plant vendors.

NON-PROCESS EQUIPMENT DATA SOURCES

TITLE:

Survey and Evaluation of System Interaction Events and Sources

SPONSOR/AUTHOR:	**NO.:**
USNRC-IE	4.7-11

INDUSTRY:	**TIME FRAME:**
Nuclear	July 1980 to December 1984
TYPE:	**FREQUENCY OF UPDATE:**
Report	None

NUMBER AND TYPE OF RECORDS: 400 occurrences of snubber failure at U.S. nuclear power plants from event reports

DATA BOUNDARY: Hydraulic and mechanical snubbers

DATA ACCESS:

Contact: Monomohan Subudhi
Brookhaven National Laboratory, Building 130, Upton, NY 11973
Phone: (516) 282-2429 (FTS 666-2429)
Report ordering address: NTIS, Springfield, VA 22161
Phone: (703) 487-4650
Report No.: NUREG/CR-3922, Vols 1&2, 12/84
Report cost: Vol.1 - $19.95, Vol.2 - $25.95

DESCRIPTION:

A review of snubber operating experience at nuclear power plants from 1980 to 1984 is given in this report. Both hydraulic and mechanical snubber types are reviewed.

The report includes an evaluation of snubber performance; operational, installation-related, and manufacturing problems are identified. General failure data with failure modes and respective frequencies is provided.

Also included are a review of pertinent vendor activities and recommendations for effective reliability and overall nuclear plant safety.

NON-PROCESS EQUIPMENT DATA SOURCES

TITLE: A Statistical Analysis of Nuclear Power Plant (Pump and Valve) Failure Rate Variability: Some Preliminary Results

SPONSOR/AUTHOR: USNRC-RES

NO.: 4.7-12

INDUSTRY: Nuclear

TIME FRAME: 1975 to 1983

TYPE: Report

FREQUENCY OF UPDATE: None

NUMBER AND TYPE OF RECORDS: All IPRDS data base records for the pumps and valves selected for analysis

DATA BOUNDARY: The set of valves and pumps selected for analysis from the IPRDS data base

DATA ACCESS:

Contact: Elizabeth Kelly, Los Alamos National Laboratory
Group S-1, MSF 600, Los Alamos, NM 87545
Phone: (505) 667-3308 (FTS 843-3308)
Report ordering address: NTIS, Springfield, VA 22161
Phone: (703) 487-4650
Report cost: $14.95

DESCRIPTION:

Los Alamos National Laboratory performed separate statistical analyses using the Failure Rate Analysis Code (FRAC) on IPRDS failure data for pumps and valves. The major purpose of the study was to determine which environmental, system, and operating factors adequately explain the variability in the failure data. The results of the pump study are documented in NUREG/CR-3650. The valve study findings are presented in NUREG/CR-4217.

In the analysis of pumps, IPRDS failure data for 60 selected pumps at four nuclear power plants were statistically analyzed using FRAC. The data cover 23 functionally different pumps, respectively, for two BWRs. Catastrophic, degraded, and incipient failure severity categories were considered for both demand-related and time-dependent failures.

For catastrophic demand-related pump failures, the variability is explained by the following factors listed in their order of importance: system application, pump driver, operating mode, reactor type, pump type, and unidentified plant-specific influences. Quantitative failure rate adjustments are provided for the effects of these factors. In the case of catastrophic time-dependent pump failures, the failure rate variability is explained by three factors: reactor type, pump driver, and unidentified plant-specific influences. Point and confidence interval failure rate estimates are provided for each selected pump by considering the influential factors. Both types of estimates represent an improvement over the estimates computed exclusively from the data on each pump. The coded IPRDS data used in the analysis is provided in an appendix. A similar treatment applies to the valve data.

NON-PROCESS EQUIPMENT DATA SOURCES

TITLE:
Investigation of Valve Failure Problems in LWR Power Plants

SPONSOR/AUTHOR:	**NO.:**
US DOE, Div. of Nuclear Power Development	4.7-13

INDUSTRY:	**TIME FRAME:**
Nuclear	February 1966 to January 1979
TYPE:	**FREQUENCY OF UPDATE:**
Report	None

NUMBER AND TYPE OF RECORDS: 195 LERs valve failures causing trips from 12/72 to 12/78, plus all valve failures for 10 stations from 2/66 to 1/79

DATA BOUNDARY: Valve-related events reported in LERs, as noted above

DATA ACCESS:

Contact: National Technical Information Service (NTIS)
Springfield, VA 22161
Phone: (703) 487-4650
Report No.: ALO-73, April 1980
Report cost: $32.95

DESCRIPTION:

The study performed by Burns and Roe (B&R) shows that valve failures constitute the component category most responsible for the shutdown of PWR and BWR plants. This investigation, contracted with SNL for DOE, identified the principal types and causes of valve failures that led to plant trips for the period from 12/72 to 12/78. The primary sources of data for the report were searches of the data base, the monthly Gray Books, Nuclear Power Experience publications, as well as discussions with utilities, valve manufacturers, and suppliers.

As the result of a cursory review of the NSIC/LER abstracts, the report investigates in greater detail the statistically most common types of valve failures. These valves include power-operated and spring-loaded relief, main steam isolation, feedwater regulator, pressurizer spray, and solenoid-operated pilot valves. Also considered are generic problems such as valve stem leakage, valve actuation malfunction, and special valves that do not require packing. Typical system flow diagrams present outline and sectional drawings of the valve and typical installation arrangements. Regulatory and code requirements as well as design responsibilities are discussed. The report provides an analysis and statistics for each valve type summarizing the utilities' and manufacturers' experience.

This study is a good reference for the construction of fault/event trees of systems that are affected by valve performance. The valve failure modes are identified, the associated mechanisms are described in detail, and preventive measures are offered.

NON-PROCESS EQUIPMENT DATA SOURCES

TITLE:

The Reliability of Emergency Diesel Generators at U.S. Nuclear Power Plants

SPONSOR/AUTHOR:

Electric Power Research Institute (EPRI)

NO.:

4.7-14

INDUSTRY:

Nuclear

TIME FRAME:

1983 to 1985

TYPE:

Report

FREQUENCY OF UPDATE:

None

NUMBER AND TYPE OF RECORDS: Number of failures and demands for 154 diesels

DATA BOUNDARY: Failure to start and failure to load and run data for Diesel Generators

DATA ACCESS:

Contact: Technical Information Center, EPRI
3412 Hillview Avenue, P.O. Box 10412, Palo Alto, CA 94303

Phone: (415) 855-2411

Report Ordering Address: Research Reports Center
P.O. Box 50490, Palo Alto, CA 94303

Phone: (415) 965-4081

Report Cost: Free to EPRI member utilties and certain other nonprofit organizations

DESCRIPTION:

EPRI's NSAC surveyed the U.S. nuclear power plant industry in order to determine diesel generator (DG) reliability. For each of 154 diesel generators, reliability data are provided. Both testing and unplanned demands and associated failures were included. However, the unplanned demands represent only about 2% of the total and have very few failures (for the 75 units and three years in the study, there were 431 start demands and 223 load-run demands with only 2 start failures and 4 load-run failures). Because of this sparsity, only the total demands and associated failures were entered into the data base.

Two failure modes were considered in the study. The first is failure to start; in addition to unplanned demands this includes both fast start tests and slow start tests. The failure to run mode includes all failures occurring from the time when load was applied to the DG until the diesel is no longer needed or until the end of the running duration required by technical specifications.

For both failure modes, terminations caused by conditions other than the DG and its immediate support systems were not counted. Conditions that invalidated tests or demands for this study include any operating errors that would not have prevented the DG from being restarted and brought to load in a few minutes without corrective maintenance; incorrect trip signals that would not have been operative in the emergency mode; and minor water or oil leaks that would not have precluded operation of the DG in an emergency.

NON-PROCESS EQUIPMENT DATA SOURCES

TITLE: EPRI Guide	
SPONSOR/AUTHOR: Electric Power Research Institute (EPRI)	**NO.:** 4.7-15
INDUSTRY: Power	**TIME FRAME:** January 1982 to August 1987
TYPE: Reports	**FREQUENCY OF UPDATE:** Varies

NUMBER AND TYPE OF RECORDS: 3000 report descriptions

DATA BOUNDARY: Advanced power, coal, electrical, nuclear, energy managment, and environment topic areas

DATA ACCESS:

Contact: Technical Information Center, EPRI
3412 Hillview Avenue, Palo Alto, CA 94303
Phone: (415) 855-2411
Report ordering address: Research Reports Center, P.O. Box 50490
Palo Alto, CA 94303
Phone: (415) 965-4081
Price: EPRI report prices depend upon the page count of each document. A subject index is available.

DESCRIPTION:

EPRI funds research on various topics dealing with the generation of electric power. Reports for most of these projects are available from the Research Reports Center free of charge to EPRI member utilities and affiliates, contributing nonmembers, U.S. utility associations, U.S. government agencies, and foreign organizations with which EPRI maintains exchange agreements. EPRI maintains a catalog of all its publications (EPRI Guide) which provides a brief synopsis for each report and can be used as a preliminary screening tool for assessment of a report's potential use.

NON-PROCESS EQUIPMENT DATA SOURCES

TITLE: Component Failure and Repair Data for Gasification-Combined Cycle Power Generation Units

SPONSOR/AUTHOR: EPRI

NO.: 4.7-16

INDUSTRY: Power

TIME FRAME: Through 1981

TYPE: Report

FREQUENCY OF UPDATE: None

NUMBER AND TYPE OF RECORDS: Failure rates and average restore times from published, analytical, and judgment data

DATA BOUNDARY: Data for 121 system/component groups from Coal Gasification Combined-Cycle Units

DATA ACCESS:

Contact: R.P. Dawkins and J.A. Derdiger
Fluor Engineers and Constructors, 2802 Kelvin St., Irvine CA 92714
Report order address: EPRI Research Reports Center
P.O. Box 50490, Palo Alto, CA 94303
Phone: (415)965-4081
Report No.: EPRI AP-2205
Report cost: Free to EPRI member utilities and affiliates

DESCRIPTION:

This report presents a set of failure rate and time-to-restore data for typical components of a coal gasification combined cycle power generation unit. The data was used to examine the reliability and availability of a generic power generation unit using risk analysis models.

The failure rates and times-to-restore developed used a variety of data sources and data construction methodologies and are presented in Section 2. The principal methodology used is a kind of failure mode analysis for each component; several principle modes of failure are analyed by characteristics including frequency of occurence, repair time, start-up time, and shut-down time. From these an average failure rate is developed and expressed as failures per million hours and mean time between failure(MTBF).

NON-PROCESS EQUIPMENT DATA SOURCES

TITLE:

Performance of Pipework in the British Sector of the North Sea

SPONSOR/AUTHOR:	**NO.:**
UKAEA	4.7-17

INDUSTRY:	**TIME FRAME:**
Offshore Oil and Natural Gas	January 1977 to January 1983

TYPE:	**FREQUENCY OF UPDATE:**
Report	None

NUMBER AND TYPE OF RECORDS: Failure rates based on 27 actual incidents from UK DOE reports

DATA BOUNDARY: Offshore oil, gas, and process fluid submarine pipelines within the UK Continental Shelf

DATA ACCESS:

Report Title: Performance of Pipework in the British Sector of the North Sea
Contact: The Editor
United Kingdom Atomic Energy Agency
Safety and Reliability Directorate, Wigshaw Lane
Culcheth, Warrington WA3 4NE
Authors: A.G. Cannon and R.C.E. Lewis
Report No.: NCSR/GR/71; 8/87

DESCRIPTION:

The aim of this project was to collect information on offshore oil, gas, and other related pipelines for the purpose of deriving reliability data. The data was gathered from various files, reports, and reference charts held at the Pipeline Unspectorate at the UK Department of Energy. Such files include construction files on new pipe, maintenance files concerning pipe work, and incident files logged by operators on any accurrence of pipeline leak or damage per the requirements of the UK DOE. The latter files provided the majority of the raw data used to compile this paper. This raw data includes incidents involving risers, valves, pig traps and main tie-in couplings (i.e. riser tie-ins) in addition to pipelines and summarises more detail which is contained in the original report.The data analysis relate to pipeline incidents only for three main categories of failure.

All reported incidents. Includes all reports of anchor, cable or trawl contact and also severe concrete coating damage.

Incidents causing shutdown. This group covers: (a) incidents where shutdown has occurred at some stage, either from damage repair or in anticipation of damage occurring, and (b) incidents where serious damage has occurred.

Incidents causing immediate shutdown. (a) This group covers all incidents where any discharge from a pipe occurred. (b) Faults in pipe laying which were not immediately found and repaired, are included.

Fault rates were found based on service, type, length, outside diameter and maximum allowed operating pressure.

NON-PROCESS EQUIPMENT DATA SOURCES

TITLE:

Reliability Analysis Center Handbooks

SPONSOR/AUTHOR:	**NO.:**
RADC	4.7-18

INDUSTRY:	**TIME FRAME:**
Government and Military	Varies
TYPE:	**FREQUENCY OF UPDATE:**
Reports	Every several years

NUMBER AND TYPE OF RECORDS: Data summaries of hundreds of records by component and environment

DATA BOUNDARY: Electronic component reliability data, i.e. microelectronic devices, high technology components

DATA ACCESS:

Contact: Steven Flint or Jeanie Lasher
Reliability Analysis Center
Rome Air Development Center
Griffiss Air Force Base, NY 13441
Phone: (315) 337-0900
Report cost: Varies per report; approx $100 per book

DESCRIPTION:

The Reliability Analysis Center (RAC) at Rome Air Development Center, Griffiss Air Force Base, maintains a comprehensive accumulation of electronic component reliability data and information representing the combined experiences of hundreds and government, industrial, and independent organizations.

Present data acquisition concentrates on microelectronic devices, high-technology components, discrete semiconductors, and nonelectronic parts. Data are solicited among all phases of device and system development, assembly, testing, and field operation. These activities are enhanced and extended through direct interaction between the RAC and the Rome Air Development Center reliability staff. Emphasis is given to failure modes and mechanisms; material, device, and process technology; quality assurance, reliability, and maintainablity practices; specifications and standards; test results; and application experience. Collected data are classified according to physical, material, design, and process control characteristics as well as applied stress environment.

RAC publications include data summaries for specific component types, such as hybrid microcircuits, small, medium and large-scale integration digital devices, linear and interface devices, digital monolithic devices, and discrete semiconductors. In addition, there are reliability and equipment maintenance data books that provide the failure and repair time data on military electronic equipment by application such as subsystem.

NON-PROCESS EQUIPMENT DATA SOURCES

TITLE: An Analysis of Reportable Incidents for Natural Gas Transmission and Gathering Lines 1970 Through June 1984

SPONSOR/AUTHOR: To American Gas Association from Battelle

NO.: 4.7-19

INDUSTRY: Natural Gas

TIME FRAME: 1970 to 1984

TYPE: Report

FREQUENCY OF UPDATE: None

NUMBER AND TYPE OF RECORDS: Several thousand incidents of service and test failures

DATA BOUNDARY: Gas pipelines

DATA ACCESS:

A.G.A. Catalog No. L51499
Contact: American Gas Association
Order Processing Department
1515 Wilson Blvd., Arlington, VA 22209
Phone: (703) 841-8400
Report cost: $25

DESCRIPTION:

This report is by Battelle Columbus Division to the Line Pipe Research Supervisory Committee of the American Gas Association. It presents an analysis of statistical data obtained from reports of leak or rupture (service) incidents and test failures in natural gas transmission and gathering lines over the 14.5 year period from 1970 through June, 1984. All gas transmission companies were required to notify the Office of Pipeline Safety Operations in the event of a "reportable" incident, as defined by the Code of Federal Regulations. The purpose of the study is to organize the reportable incident data into a meaningful format from which the safety record of the industry can be assessed.

NON-PROCESS EQUIPMENT DATA SOURCES

TITLE:

Severities of Transportation Accidents

SPONSOR/AUTHOR: R. K. Clarke, J. T. Foley, W. F. Hartman, D. W. Larson, Sandia National Laboratories

NO.: 4.7-20

INDUSTRY:

Automotive, Airline, Truck

TIME FRAME:

Late 1960's to early 1970's

TYPE:

Report

FREQUENCY OF UPDATE:

None

NUMBER AND TYPE OF RECORDS: Graphs and tables giving probability of accidents having certain severities

DATA BOUNDARY: US transportation industry

DATA ACCESS:

Report No.: SLA-74-0001

Report ordering address: Engineering Analysis Dept.
Sandia National Labs
Albuquerque, NM 87115

Report accessibility: unlimited release

DESCRIPTION:

This report documents the development of data on the severity as well as the frequency of accidents involving truck, rail, and air transport. Volume 1 includes a summary giving the probability of occurrence of accidents as a function of accident severity. Subsequent Volumes give supporting data, calculations and analysis.

NON-PROCESS EQUIPMENT DATA SOURCES

TITLE:
Pressure Vessel Failure Statistics and Probabilities

SPONSOR/AUTHOR: J. R. Engel, AEC Advisory Committee on Reactor Safeguards

NO.: 4.7-21

INDUSTRY: Nuclear

TIME FRAME: Through 1971

TYPE: Journal Article

FREQUENCY OF UPDATE: None

NUMBER AND TYPE OF RECORDS: 4 tables containing failure data for vessels

DATA BOUNDARY: Primarily concerned with boiler failures

DATA ACCESS:

Contact: Nuclear Safety, Vol. 15, No. 4, July - August 1974

DESCRIPTION:

This report summarizes data on non-nuclear pressure vessel failures in order to develop data which could be applied to the nuclear power industry. Tables 3 through 6 present summaries of vessel failures and failure rates.

NON-PROCESS EQUIPMENT DATA SOURCES

TITLE:
Characteristics of Pipe System Failures in Light Water Reactors

SPONSOR/AUTHOR:	**NO.:**
EPRI	4.7-22

INDUSTRY:	**TIME FRAME:**
Nuclear	Unknown (prior to 1977)
TYPE:	**FREQUENCY OF UPDATE:**
Report	None planned

NUMBER AND TYPE OF RECORDS: Approximately 100 records of pipe failure rates in a wide variety of failure modes.

DATA BOUNDARY: Nuclear Power Plant Piping

DATA ACCESS:

Report No.: EPRI NP-438
Contact: Research Project 705-1
Report ordering address: Electric Power Research Institute
Research Reports Center
P.O. Box 50490, Palo Alto, CA 94304
Phone No.: (415) 965-4081
Report cost: $25.00

DESCRIPTION:

This report is a statistical description of pipe system failures. The characteristics of these failures have been derived from reports submitted by the utilities to the Nuclear Regulatory Commission. The bulk of the data is from Licensee Event Reports supplemented as necessary by plant outage and maintenance data.

NON-PROCESS EQUIPMENT DATA SOURCES

TITLE:
Reliability of Emergency AC Power Systems

SPONSOR/AUTHOR:
USNRC-NRR

NO.:
4.7-23

INDUSTRY:
Nuclear

TIME FRAME:
January 1976 through December 1980

TYPE:
Report

FREQUENCY OF UPDATE:
One update in 1985

NUMBER AND TYPE OF RECORDS: 900 occurrences of diesel generator failures at U.S. nuclear power plants

DATA BOUNDARY: Diesel generator performance data for 18 different nuclear power plants

DATA ACCESS:

Contact: Ronald E. Battle, Oak Ridge National Laboratory
Building 3500, MS-8, Oak Ridge, TN 37831
Phone: 615-574-5531 (FTS 624-5531)
Report ordering address: GPO Sales Program, Division of Technical Information and Document Control
Report cost: $8.00 Report accessibility: No Restrictions
Report No.: NUREG/CR-2989

DESCRIPTION:

Station blackout, or loss of all AC power, has been identified in many PRAs as a major contributor to risk. This is because of the disablement of all normal and most emergency cooling systems that occur during loss of AC power. In addition, most engineered safety feature systems that would contain radioactive material given a nuclear accident are disabled as a result of station blackout. The seriousness of these consequences provided the major motivation for this study. Specific plants were selected to estimate onsite AC power system reliability based on the most realistic data available, but with the intent of using the data as representative figures for any plant with the design and operational features identified in this report.

The sources of data for this report were: (a) Abstracts of LERs, (b) LOCA data submitted to the USNRC by licensees in response to a questionnaire associated with NUREG-0737, (c) Diesel generator data submitted to the USNRC in response to a questionnaire prepared as part of the study.

The bulk of the information in the report is included in a 317-page appendix that contains systems descriptions, station blackout fault trees, diesel generator historical data, and diesel generator common cause failure analysis results for 18 different nuclear power plants. Tables and graphs are well organized and present data correlated to each plant studied. The study also contains conclusions and recommendations for improving reliability.

4.8 INDEX OF PROBABILISTIC RISK ASSESSMENTS

NO.	TITLE	INDUSTRY	NO. & TYPE OF RECORDS	DATA BOUNDARY	PAGE
4.8-1	Big Rock Point Probabilistic Risk Assessment	Nuclear	30 component failure rates; 20 maintenance unavailabilities	Typical nuclear plant PRA components and failure modes	117.
4.8-2	Connecticut Yankee Probabilistic Safety Study	Nuclear	Failure and maintenance data using experience to update generic rates	Pumps, MOVs, batteries, chargers, inverters, motors, and diesels	118.
4.8-3	Indian Point Units 2 and 3	Nuclear	Generic data set based on plant and industry failure reports over a ten-year span	Failure rates for standard nuclear PRA components and failure modes	119.
4.8-4	Probabilistic Risk Assessment, Limerick Generating Station	Nuclear	Failure rates and modes from plant records, other plant reports, WASH-1400 data	Standard nuclear PRA components (pumps, valves, diesels, instruments)	120.
4.8-5	Interim Reliability Evaluation Program: Analysis of the Millstone Point 1 Nuclear Power Plant	Nuclear	Component failure and maintenance data from plant records for a 13.5 year span	Primarily focuses on pumps, motor-operated valves, and breakers	121.
4.8-6	Oconee-3 PRA A Probabilistic Risk Assessment of Oconee Unit 3	Nuclear	Plant-specific data from failure records, test reports, and operating logs for a nine year span	Equipment includes pumps, valves, batteries, chargers, and breakers	122.
4.8-7	Yankee Nuclear Power Station Probabilistic Safety Study	Nuclear	Failure rates based on records from various in-plant information files for a 22-year span	Mechanical and electrical component types, especially pipes and tubes	123.
4.8-8	Zion Probabilistic Safety Study	Nuclear	Generic and plant-specific failure rate data	Component level failure rates for typical PRA data set	124.
4.8-9	Reactor Safety Study: An Assessment of Accident Risk in U.S. Commercial Nuclear Power Plants (WASH-1400)	Nuclear	Failure data from eight PWRs and nine BWRs for 1972 and from varied non-nuclear industry sources	Pumps, valves, electrical, some instrumentation	125.

PROBABILISTIC RISK ASSESSMENTS

TITLE:

Big Rock Point Probabilistic Risk Assessment

SPONSOR/AUTHOR:	**NO.:**
Consumers Power Co.	4.8-1

INDUSTRY:	**TIME FRAME:**
Nuclear	1970 to 1979

TYPE:	**FREQUENCY OF UPDATE:**
Report	None

NUMBER AND TYPE OF RECORDS: 30 component failure rates; 20 maintenance unavailabilities

DATA BOUNDARY: Typical nuclear plant PRA components and failure modes

DATA ACCESS:

Data Accessibility: PRA Documents generally have a limited distribution and comprise several volumes of reports. The data may be available in government or national laboratory libiaries but access is usually restricted to the utility and contractors who performed the analysis.

DESCRIPTION:

Appendix III of this report provides a detailed description of the reliability data used in event tree and fault tree quantification. Because of its extensive operating experience and the uniqueness of the BRP design, BRP plant-specific data was used whenever possible. Plant-specific data sources included plant maintenance orders, control room log books, surveillance tests, LERs, event reports, deviation reports, plant review committee meeting minutes, and USNRC correspondence. The plant-specific data used spanned the period from 1970 to 1979. Data before 1970 did not include maintenance orders or surveillance tests and therefore were excluded. The plant-specific data collected for BRP is presented in detail in Appendix XIII. Table III-4 summarizes 30 plant-specific component failure rates and Table II-06 contains plant-specific maintenance unavailabilities for 20 components. These tables are a summary of the BRP component failure and maintenance outages.

The generic data sources used in the BRP data base originate from the nuclear industry.

A systematic approach was undertaken for the BRP PRA to identify all potential sources of common mode failure. The first step in the treatment of common mode failures was a compilation of a detailed list of common mode initiators. To achieve this, available literature on common mode failure analysis was reviewed. The next step was to qualitatively assess the potential effects of these initiators on BRP systems. The initiator categories and the systems selected for examination are presented in Table VI.1 of the BRP PRA.

PROBABILISTIC RISK ASSESSMENTS

TITLE:
Connecticut Yankee Probabilistic Safety Study

SPONSOR/AUTHOR:
Northeast Utilities

NO.:
4.8-2

INDUSTRY:
Nuclear

TIME FRAME:
1976 to 1986

TYPE:
Report

FREQUENCY OF UPDATE:
None

NUMBER AND TYPE OF RECORDS: Failure and maintenance data using experience to update generic rates.

DATA BOUNDARY: Pumps, MOVs, batteries, chargers, inverters, motors, and diesels

DATA ACCESS:

Data Accessibility: PRA Documents generally have a limited distribution and comprise several volumes of reports. The data may be available in government or national laboratory libiaries but access is usually restricted to the utility and contractors who performed the analysis.

DESCRIPTION:

The Connecticut Yankee (Haddam Neck Plant) PRA contains component failure and maintenance unavailability data, and initiating event frequency data including typical PWR anticipated operational occurrences, LOSP, and RCP seal failures. Section 4.1.1., "Component Data Collection," describes the process used to gather component failure history, demand history, and run time experience over a 10-year period. Included in the process was the use of the Baseline Events Analysis Reliability Data System (BEARDS), a proprietary Northeast Utilities data base that includes failure and maintenance reports. Section 4.1.2, "Component Reliability Analysis," shows the updated component failure data. The results reflect a Bayesian update of WASH-1400, IEEE-500, and Westighouse Nuclear Technology Division means and variances, using plant-specific experience where available. Failure-on-demand rates were modeled using beta-distributed priors. Hourly failure rates were modeled using gamma-distributed priors. The pump data base is broken down into pump types (each is treated separately) and failure modes (start versus run). Limited motor-operated valve data were analyzed for critical, infrequently tested valves inside the containment. No plant-specific breaker data was obtained. Good plant-specific data exist for batteries, chargers, inverters, motors, and diesels.

PROBABILISTIC RISK ASSESSMENTS

TITLE:
Indian Point Units 2 and 3

SPONSOR/AUTHOR:
Consolidated Edison and New York Power Authority

NO.:
4.8-3

INDUSTRY:
Nuclear

TIME FRAME:
May 1973 to December 1979

TYPE:
Report

FREQUENCY OF UPDATE:
None

NUMBER AND TYPE OF RECORDS: Generic data set based on plant and industry failure reports over a ten-year span

DATA BOUNDARY: Failure rates for standard nuclear PRA components and failure modes

DATA ACCESS:

Data Accessibility: PRA Documents generally have a limited distribution and comprise several volumes of reports. The data may be available in government or national laboratory libiaries but access is usually restricted to the utility and contractors who performed the analysis.

DESCRIPTION:

The review of the data portion of the Indian Point 2 (IP2) and 3 (IP3) PRA (a 1982 internal document prepared by Consolidated Edison and the New York Power Authority) is confined to the plant-specific and generic component failure and service hour data sections because these were the only segments available to the reviewers. The LERs produced during a ten-year span of IP2's operation were evaluated to determine their applicability to the PRA data needs. It was eventually decided to use only the LERs generated after IP2 became critical (from May 23, 1973 to December 31, 1979) for the component data base development, based on the availability of failure event information and more uniform operability, testing, and reporting criteria.

Opening segments of the IP2 PRA data analysis section describe the definitions of terms and concepts employed, the assumptions made, and limitations recognized during the data base construction. A set of 39 plant-specific component failure mode summaries established the basis for component service hour determinations, the number of failures, and the test data source for each failure mode given for each component. Generic data from WASH-1400, IEEE Std 500, and the LER data summaries on valves, pumps, and diesels were combined with plant-specific failure data to produce "updated" failure information. All the IP2 specialized component hardware failure data, both generic and updated, are contained in Table 1.5.1-4 (IP3: 1.6.1-4). This table contains (by system, component, and failure mode) plant-specific data on the number of failures and service hours or demands. For some components, it was determined that specifications of the system was warranted because of its impact on the data values.

PROBABILISTIC RISK ASSESSMENTS

TITLE:

Probabilistic Risk Assessment, Limerick Generating Station

SPONSOR/AUTHOR:	**NO.:**
Philadelphia Electric Co.	4.8-4

INDUSTRY:	**TIME FRAME:**
Nuclear	1976 to 1982
TYPE:	**FREQUENCY OF UPDATE:**
Report	None

NUMBER AND TYPE OF RECORDS: Failure rates and modes from plant records, other plant reports, WASH-1400 data

DATA BOUNDARY: Standard nuclear PRA components (pumps, valves, diesels, instruments)

DATA ACCESS:

Data Accessibility: PRA Documents generally have a limited distribution and comprise several volumes of reports. The data may be available in government or national laboratory libiaries but access is usually restricted to the utility and contractors who performed the analysis.

DESCRIPTION:

The purpose of this analysis was to assess the risk of operating Limerick Station, specifically with regard to its location near a high population density area. These risks were evaluated to determine whether they represent a disproportionately high segment of the total societal risk from postulated nuclear reactor incidents.

The Limerick analysis accounted for a revised list of incident initiators based on the Limerick plant design and a more detailed analytical modeling of event sequences following each incident initiator. Plant-design-specific and site-specific data were also included in the analysis of the Limerick Mark II containment and in the meterology and demography imput to the evaluation of incident consequences.

The component failure rate data used as input to the fault tree model came from four basic sources: plant records from Peach Bottom (a plant of similar design to Limerick), actual nuclear plant operating experience data as reported in LERs (to produce demand failure rates evaluated for pumps, diesels, and valves), General Electric BWR operating experience data on a wide variety of components (e.g., safety relief SRV valves, level sensors containment pressure sensors), and WASH-1400 assessed median values.

PROBABILISTIC RISK ASSESSMENTS

TITLE: Interim Reliability Evaluation Program: Analysis of the Millstone Point 1 Nuclear Power Plant

SPONSOR/AUTHOR: Northeast Utilities

NO.: 4.8-5

INDUSTRY: Nuclear

TIME FRAME: 1972 to 1985

TYPE: Report

FREQUENCY OF UPDATE: Continuous

NUMBER AND TYPE OF RECORDS: Component failure and maintenance data from plant records for a 13.5 year span

DATA BOUNDARY: Primarily focuses on pumps, motor-operated valves, and breakers.

DATA ACCESS:

Data Accessibility: PRA Documents generally have a limited distribution and comprise several volumes of reports. The data may be available in government or national laboratory libiaries but access is usually restricted to the utility and contractors who performed the analysis.

DESCRIPTION:

The Millstone Unit 1 PRA contains component failure and maintenance unavailability data, and initiating event frequency data, including typical BWR anticipated operational occurrences and LOSP.

The pump data is broken down into pump types (each is treated separately) and failure modes (start versus run). The MOV data is separated into MOVs inside the drywell versus those outside the drywell (statistically significant differences exist in observed failure rates). The MOV data also reflects failure to open versus failure to close. The large electrical breaker (4160 V, 480 V) data shows significant differences from both the WASH-1400 and IEEE Std.500 data, i.e., 3 failures in 34,333 demands for 4160 V breakers and 6 failures in 11,238 demands for 480 V breakers.

Maintenance data are treated by computing an average maintenance unavailability for each component type or system and fitting the data to a beta distribution. This is because maintenance outages are logged on a system basis in many cases.

Event frequency data were developed from a detailed review of plant trip reports and shift supervisor's logbook entries. The first two years of plant experience were discarded as they appear to represent experience typical of early plant operation and tests that are not typical of operation in later years. The plant experience was used to perform a Bayesian update of EPRI NP-2230 reactor trip experience.

PROBABILISTIC RISK ASSESSMENTS

TITLE:
OCONEE-3 PRA A Probabilistic Risk Assessment of Oconee Unit 3

SPONSOR/AUTHOR:	**NO.:**
EPRI & Duke Power	4.8-6

INDUSTRY:	**TIME FRAME:**
Nuclear	1975 to 1984
TYPE:	**FREQUENCY OF UPDATE:**
Report	None

NUMBER AND TYPE OF RECORDS: Plant-specific data from failure records, test reports, and operating logs for a nine year span

DATA BOUNDARY: Equipment includes pumps, valves, batteries, chargers, and breakers

DATA ACCESS:

Report ordering address: Electric Power Research Institute (EPRI)
Research Reports Center, P.O. Box 50490
Palo Alto, CA 94303

Phone: (415) 965-4081
Report No.: NSAC-60, Volumes 1-5, June 1984
Report Cost: $350.00 for all 5 volumes

DESCRIPTION:

The Oconee Unit 3 PRA contains plant-specific raw data for valves; pumps; reactor building cooling units; isolating diode assemblies; instrument inverters; four sizes of transformers; panel boards; low and high voltage busses; buswork; DC, low voltage, and high voltage circuit breakers; batteries; battery chargers; and hydro-driven generators. Nearly all the data is system-specific except for the valve data. For motor-operated valves that fail to operate, data for the condenser circulating water system is separately listed.

The failure data for these rates is obtained from maintenance work requests supplemented by incidence reports and Licensee Event Reports from the 1975-1980 time period. The work requests provide a complete history of all repairs performed at Oconee. They are not restricted to safety-related systems, they are written during all modes of unit operation, and they are not produced in response to licensing-based criteria.

Periodic test reports, control room operating logs, and piping and instrumentation diagrams were used to compute the numbers of demands. Operating hours were based on run hour logs for the motor-driven pumps and cooling units; a review of normal plant operating procedures, system lineups, and periodic test records provided component service hours for other component/failure mode combinations. An appendix to the PRA contains a data summary table for each failure rate estimate describing its basis.

PROBABILISTIC RISK ASSESSMENTS

TITLE:
Yankee Nuclear Power Station Probabilistic Safety Study

SPONSOR/AUTHOR:
Yankee Atomic Electric Co.

NO.:
4.8-7

INDUSTRY:
Nuclear

TIME FRAME:
1961 to 1983

TYPE:
Report

FREQUENCY OF UPDATE:
None

NUMBER AND TYPE OF RECORDS: Failure rates based on records from various in-plant information files for a 22-year span

DATA BOUNDARY: Mechanical and electrical component types, especially pipes and tubes

DATA ACCESS:

Data Accessibility: PRA Documents generally have a limited distribution and comprise several volumes of reports. The data may be available in government or national laboratory libiaries but access is usually restricted to the utility and contractors who performed the analysis.

DESCRIPTION:

The Yankee Nuclear Power Station was designed during the 1950s, and was first licensed for operation in 1961 by the U.S. Atomic Energy Commission. Because of its age, an extensive history of plant experience was available for use for the comprehensive risk analysis performed in 1983. Plant information records, maintenance department information (such as surveillance schedules and machinery history cards), instrument and controls department item identification index, and reactor engineering department operating data report and statistics all provided valuable data resources for the PRA. LERs and NPRDS data were also used to reduce the data reduction workload to a manageable effort. Special generic data studies on component-specific issues were accessed to supplement this plant-specific information.

The major data areas addressed are: initiating events, sequence data, top event data components, and human error data. An alphabetical presentation of component types (mechanical, then electrical), subtypes where deemed necessary, and failure rate data in terms of mean, median, range factor, and variance values are logged in Table 7-2. Special consideration was given in the PRA to piping and tube failure rates; therefore, mean and variance values are cited for small, intermediate, and large LOCAs and secondary system piping failures in Table 7-7.

The component hardware data were well-based because they are derived from 22 years of records. It is also useful because of the structure and depth of the presentation; for example, the inclusion of data on pumps in different systems (emergency feedwater, condensate, service water).

PROBABILISTIC RISK ASSESSMENTS

TITLE:
Zion Probabilistic Safety Study

SPONSOR/AUTHOR:
Commonwealth Edison Co.

NO.:
4.8-8

INDUSTRY:
Nuclear

TIME FRAME:
1975 to 1981

TYPE:
Report

FREQUENCY OF UPDATE:
None

NUMBER AND TYPE OF RECORDS: Generic and plant-specific failure rate data

DATA BOUNDARY: Component level failure rates for typical PRA data set

DATA ACCESS:

Data Accessibility: PRA Documents generally have a limited distribution and comprise several volumes of reports. The data may be available in government or national laboratory libiaries but access is usually restricted to the utility and contractors who performed the analysis.

DESCRIPTION:

The detailed risk assessment conducted for the Zion station considered both Units 1 and 2. A comprehensive data base, covering topics similar to those dealt with in the Shoreham and Oconee Unit 3 PRAs, is discussed and presented in PRA Section II.4.4 of the report on Data Base Development.

The Zion PRA data base includes generic, plant-specific, and combined "updated" component failure data, maintenance frequencies for components, initiating event data, human error rates, and component operability, test and service hour data. A nine-page component failure data table specifies mean values and 60% confidence interval error factors for generic data and updated mean values and variances for particular component types and failure modes. Most of the component failure rates were applicable to all systems, but exceptions are noted in some cases. Tables with maintenance frequency mean and variance values for selected components; tables with initiating event occurrence probability mean, median, and 90% confidence bound values; and fluid systems unavailability values are among the Zion PRA Data Base tables with features. A series of graphs shows the distribution of probability density versus occurrences per year for each initiating event; for example, loss of Reactor Cooling System flow, core power excursion, and turbine trip. A System Description and Analysis Summary section provides brief descriptions of the safety systems essential to core damage prevention.

PROBABILISTIC RISK ASSESSMENTS

TITLE: Reactor Safety Study: An Assessment of Accident Risk in U.S. Commercial Nuclear Power Plants (WASH-1400)

SPONSOR/AUTHOR: USNRC

NO.: 4.8-9

INDUSTRY: Nuclear

TIME FRAME: Through 1974

TYPE: Report

FREQUENCY OF UPDATE: None

NUMBER AND TYPE OF RECORDS: Failure data from eight PWRs and nine BWRs for 1972 and from varied non-nuclear industry sources.

DATA BOUNDARY: Pumps, valves, electrical, some instrumentation

DATA ACCESS:

Report ordering: National Technical Information Service (NTIS)
Springfield, VA 22161
Phone: (703) 487-4650
Report cost: $68 for all volumes
Report accessibility: No restrictions

DESCRIPTION:

The Reactor Safety Study (WASH-1400) was published by the USNRC in 1975 to set down a methodology for assessing nuclear plant reliability and risk. Of particular interest to the data analyst are Appendix III, "Failure Data," and Appendix IV, "Common Mode Failures."

Appendix III contains failure rate estimates for various generic types of mechanical and electrical equipment. Included are listings of failure rates with range estimates for specified component failure modes, demand probabilities, and times to maintain repair. It also contains some discussion on such special topics as human errors, aircraft crash probabilities, loss of electric power, and pipe breaks. Appendix III contains a great deal of general information of use to analysts on the methodology of data assessment for PRA.

Appendix IV contains a thorough discussion of quantification techniques and engineering studies of common mode failures. Large LOCA, small LOCA, and transient sequences are considered.

WASH-1400 is a fundamental document for PRA methodology. The data appendixes contain a great deal of useful information on methods of data assessment. A large number of sources for data are considered, and very general failure rate estimates will produce only gross approximations. Since the advent of data collection schemes across and within plants, the WASH-1400 data are solely useful as a constituent to a data aggregation process or as widely bounded figures that provide a basis for comparison.

5
CCPS Generic Failure Rate Data Base

This chapter contains tables of generic equipment failure rate data for some of the CPI equipment types listed in Appendix A, the CCPS Taxonomy, or in Appendix B, the Equipment Index. Section 5.1 on data selection explains how data were selected from resources and lists which resources in Chapter 4 were used to provide data.

In certain cases, more than one data point was available for a given data cell table in the CCPS Taxonomy. When several data points were considered appropriate and applicable to process equipment, the data were combined through a computer-aided aggregation process. The aggregation process is described in Section 5.2.

Section 5.3, Data Table Presentation, illustrates the format used for data tables and explains the type of information contained. Data tables have been presented only for those data cells where data existed at that level in the taxonomy. These are listed by taxonomy number in the Data Cell Index, Table 5.2.

Section 5.4 describes the use of the CCPS Generic Failure Rate Data Base. Lastly, Section 5.5 contains tables of data in the Generic Failure Rate Data Base, organized by the numbers used to structure the CCPS Taxonomy.

5.1 Data Selection

SAIC provided much of the data used in this book from its proprietary files of previously analyzed and selected information. Since these data were primarily from the nuclear power industry, a literature search and industry survey described in Chapter 4 were conducted to locate other sources of data specific to the process equipment types in the CCPS Taxonomy. Candidate data resources identified through this effort were reviewed, and the appropriate ones were selected. Applicable failure rate data were extracted from them for the CCPS Generic Failure Rate Data Base. The resources that provided failure information are listed in Table 5.1 with data reference numbers used in the data tables to show where the data originated.

The selection of data from the resources available required decisions about the acceptable quality of the data and applicability of the data to the CCPS Taxonomy. Data from a resource was rejected by SAIC and the CCPS Subcommittee when:

- it duplicated data in another resource (to avoid double counting occurring through use of "incestuous" data);
- it did not clearly relate to a taxonomy level;

TABLE 5.1
Resources Used for Data Tables

Data Reference No.	Data Resource Title	Chapter 4 Resource No.
1.	Development of an Improved Liquefied Natural Gas Plant Failure Rate Data Base.	4.3-2
2.	Pressure Vessel Reliability.	4.4-1
3.	Some Data on the Reliability of Pressure Equipment in the Chemical Plant Environment.	4.4-3
4.	Some Data on the Reliability of Instruments in the Chemical Plant Environment	4.4-4
5.	Failure and Maintenance Data Analysis at a Petrochemical Plant.	4.4-5
6.	Hazardous Waste Tank Failure.	4.5-1
7.	Reliability Data Book for Components in Swedish Nuclear Power Plants.	4.6-6
8.*	SAIC Proprietary Data Set containing data from:	4.6-10
8.1	The In-Plant Reliability Data Base for Nuclear Power Plant Components.	4.6-11
8.2	IEEE Standard 500-1984.	4.6-12
8.3	Generic Data Base for Data and Models Chapter of the National Reliability Evaluation Program Guide (NREP).	4.6-13
8.4	Offshore Reliability Data Handbook (OREDA).	4.6-14
8.5	RADC Non-Electronic Reliability Notebook.	4.6-15
8.6	Reliability Prediction of Electronic Equipment (Military Handbook 217E).	4.6-16
8.7	Data Summaries of Licensee Event Reports at U.S. Commercial Nuclear Power Plants (Various Components).	4.7-8
8.8	Reliability of Emergency Diesel Generators at U.S. Nuclear Power Plants.	4.7-14
8.9	Big Rock Point Probabilistic Risk Assessment.	4.8-1
8.10	Indian Point Units 2 and 3 Probabilistic Risk Assessment.	4.8-3
8.11	Interim Reliability Evaluation Program: Analysis of the Millstone Point 1 Nuclear Power Plant Assessment.	4.8-5
8.12	Oconee-3 PRA: A Probabilistic Risk Assessment of Oconee Unit 3.	4.8-6
8.13	Yankee Nuclear Power Station Probabilistic Safety Study.	4.8-7
8.14	Zion Probabilistic Safety Study.	4.8-8
8.15	Reactor Safety Study: An Assessment of Accident Risk in U.S. Commercial Nuclear Power Plants (WASH-1400).	4.8-9
9.	An Analysis of Reportable Incidents for Natural Gas Transmission and Gathering Lines—1970 through June 1984.	4.7-19
10.	Pressure Vessel Failure Statistics and Probabilities.	4.7-21

*Note: SAIC has selected some data from resources 8.1 through 8.15 to construct its proprietary data files for use in performing PRAs. Relevant data from these files was used to construct the CCPS Generic Failure Rate Data Base. Accordingly, all usable data points contained in the resources used by SAIC may not be in the Data Tables in this book.

- it was considered to be based on insufficient data (zero failures, less than 100 demands, or less than 100 exposure hours); or
- it was of unacceptable quality.

It should be noted that data were not rejected through consideration of upper or lower bounds. These limits for the input data included a variety of assumed and calculated limits using various levels of confidence.

The quality of the existing SAIC data base was considered acceptable. Acceptability of data from CPI data resources was determined on a resource by resource and data point by data point basis. The data that were available for this book were generally derived from information that resulted from analysis and statistical treatment of original plant operating and repair records; it is unusual for published data resources to contain "raw" data. The quality of published data, therefore, depends to some extent upon the expertise and judgment of the analyst who compiled the data. By reading the text of the resource to understand where the data originated and how they were analyzed and treated, a decision was made to accept or reject the data. The experience and judgment of the SAIC data analyst who made the initial data selection played an important part in this process.

If the data quality was acceptable, they were then evaluated for their relevance and fit to the CCPS Taxonomy. The data in the SAIC data base were fitted to taxonomy levels that best correlated with nuclear plant equipment and operational environments. CPI resources were reread thoroughly to understand the equipment subtypes, operating modes, and process severities represented by the data points and to identify as many relevant taxonomy levels as possible. SAIC data analysts made preliminary judgments on the applicability of data points to taxonomy levels and on the quality of the data. The majority of the data applied to high taxonomy levels (x.x) and a smaller amount was applicable to lower levels (x.x.x.x). The data were assigned to the lowest level possible.

High levels of equipment description were generally easy to distinguish, such as AC motors from DC motors or motor-driven pumps from turbine-driven pumps. Subsequent levels became increasingly difficult to specify. For example, Lees' data resource for instruments in the CPI (Data resource 4.4-4) provides a breakdown of subtypes under the category "Analyzer." For some of Lees' categories, correspondence with the CCPS Taxonomy was clear, such as "pH Meter" with taxonomy level 2.1.1.2.7, pH analyzer. However, there were no specific taxonomy levels for Lees' data on O_2 and CO_2 analyzers. These data points were, therefore, combined with data from all analyzer sublevels to generate the data presented at the higher taxonomy level, 2.1.1, Analyzers. In these cases, only data values for the upper and lower bounds were presented, since an aggregated mean value would be of questionable validity.

Experienced SAIC analysts made the initial decisions on acceptability and assignment of data to cells. These were reviewed by the subcommittee. This process resulted in one or more sets of data points for each of a number of cells at various taxonomy levels.

5.2 Data Treatment

Failure rate data selected for the CCPS Generic Failure Rate Data Base were handled using dBase III Data Management in conjunction with the Computerized Aggregation of Reliability Parameters (CARP) developed by SAIC. CARP, designed to be used by reliability data analysts, is also written in dBase III and is compiled to make it a stand-alone program that can be run on an IBM-compatible PC. SAIC's program has several basic capabilities: data storage, identification, and handling; aggregation of generic data; calculation of uncertainty bounds (5th and 95th percentiles) for actual component failure rate statistics; distribution fitting; and report writing.

The CARP input form, Figure 5.1, contains fields for the following data:

- Equipment type, subtypes
- Operating mode
- Failure mode
- Data resource
  - Source/name
  - Component ID
  - Failure mode
  - Number of failures
  - Number of demands
  - Exposure time
  - Mean
  - Median
  - Error factor or bounds (5th or 95th)
- Data point (mean or median)
- Error factor or error bounds
- Data point weighting factors (discussed in greater detail below)
- Comments—text field of 100 characters allowing notes on the data or the data resource.

CARP's data files can also be printed out for review and quality assurance checks of the data.

The data points from SAIC's data base were already stored in CARP files. Data extracted from the CPI data resources were also entered into CARP files for storage and organization of the data points by their relevant taxonomy levels.

When more than one data point was selected for a given taxonomy level, CARP was used to combine the data points statistically. CARP allows the data points being aggregated to be weighted either equally or unequally. Unequal weighting can be used to address tolerance or confidence issues discussed in Chapter 2. Data points also can be given larger or smaller weights to reflect the data analyst's knowledge of the data pool. All of the data points used to develop the data tables in this book were weighted equally.

Standard statistical aggregation of several data points tends to produce a representative central tendency (mean, median, or mode). However, it also produces a nonrepresentative spread of the distributions (variance, error factor). The aggregate spread must reflect both the magnitude (variance) and the shape (skewness, tails, bi-modality, etc.) of the differences between data points. Many approaches are not entirely appropriate for handling generic data because the variance calculated by standard statistical techniques tends to shrink as more sources are considered and more information incorporated. During aggregation, the incorporation of additional data is treated as an increased sample size or number of experiments, which reduces the calculated variance. Unfortunately, in assembling generic data, as the sample size increases, less homogeneous data results, and the dispersion usually increases. Consequently, the spread between the upper and lower bounds for the aggregated data set is a reflection of the nonhomogeneity of the input data

```
CLASSIFICATION      COMPONENT TYPE          SUBTYPE1                          SUBTYPE2                              FAILURE MODE                              OPERATING MODE
SYSTEM ID/CODE                      SOURCE/PAGE      SOURCE FAILURE MODE                                UNIT ID COMP ID    NSSS/CODE RXT    RXA        INDUSTRY GEOGRAPHIC
#FAIL      DEMAND     EXP TIME      UNIT MEAN       MEDIAN     5TH        95TH        ERROR FACTOR
=======================================================================================================================================================================
MECHANICAL          ,PUMP                     ,TURBINE DRIVEN                                                         ,FAILS TO RUN                                ,RUNNING
FEEDWATER                     /CHA    ,NUREG 2886/111      ,FAILURE TO RUN                                ,PLANT 3,           ,GE    /C  ,BWR    ,COM          ,N        ,US
  13          0         160000      H    8.13E-05  7.75E-05  4.65E-05  1.29E-04    2.00
-----------------------------------------------------------------------------------------------------------------------------------------------------------------------
MECHANICAL          ,PUMP                     ,TURBINE DRIVEN               ,                                         ,FAILS TO RUN                                ,RUNNING
STEAM GENERATOR FEED.     /       ,NUREG 2886/67       ,FAILURE TO RUN                                ,PLANT 1,           ,WEST /E   ,PWR    ,COM          ,N        ,US
   8          0          48000      H    1.67E-04  1.54E-04  7.84E-05  3.01E-04    2.00
-----------------------------------------------------------------------------------------------------------------------------------------------------------------------
MECHANICAL          ,PUMP                     ,TURBINE DRIVEN               ,                                         ,FAILS TO RUN                                ,STANDBY
                          /       ,NUREG 1205/         ,DOES NOT OPERATE WITHOUT COMMAND FAULTS            ,       ,          ,GE    /C  ,BWR    ,COM          ,N        ,US
   0          0              0      D    2.90E-02  2.88E-02  2.38E-02  3.48E-02    1.00
-----------------------------------------------------------------------------------------------------------------------------------------------------------------------
MECHANICAL          ,PUMP                     ,TURBINE DRIVEN               ,                                         ,FAILS TO RUN                                ,STANDBY
                          /       ,NUREG 1205/         ,DOES NOT OPERATE WITHOUT COMMAND FAULTS            ,       ,          ,GE    /C  ,BWR    ,COM          ,N        ,US
   0          0              0      H    5.70E-05  5.66E-05  4.69E-05  6.84E-05    1.00
-----------------------------------------------------------------------------------------------------------------------------------------------------------------------
MECHANICAL          ,PUMP                     ,TURBINE DRIVEN               ,                                         ,FAILS TO RUN                                ,STANDBY
                          /       ,NUREG 1205/         ,DOES NOT OPERATE WITHOUT COMMAND FAULTS            ,       ,          ,CE    /B  ,PWR    ,COM          ,N        ,US
   0          0              0      H    5.30E-05  5.11E-05  3.29E-05  7.95E-05    2.00
-----------------------------------------------------------------------------------------------------------------------------------------------------------------------
MECHANICAL          ,PUMP                     ,TURBINE DRIVEN               ,                                         ,FAILS TO RUN                                ,STANDBY
                          /       ,NUREG 1205/         ,DOES NOT OPERATE WITHOUT COMMAND FAULTS            ,       ,          ,      /   ,PWR    ,COM          ,N        ,US
   0          0              0      H    3.30E-05  3.28E-05  2.71E-05  3.96E-05    1.00
-----------------------------------------------------------------------------------------------------------------------------------------------------------------------
MECHANICAL          ,PUMP                     ,TURBINE DRIVEN               ,                                         ,FAILS TO RUN                                ,STANDBY
                          /       ,NUREG 1205/         ,DOES NOT OPERATE WITHOUT COMMAND FAULTS            ,       ,          ,      /   ,PWR    ,COM          ,N        ,US
   0          0              0      D    1.60E-02  1.59E-02  1.32E-02  1.92E-02    1.00
-----------------------------------------------------------------------------------------------------------------------------------------------------------------------
MECHANICAL          ,PUMP                     ,TURBINE DRIVEN               ,                                         ,FAILS TO RUN                                ,STANDBY
                          /       ,NUREG 1205/         ,DOES NOT OPERATE WITHOUT COMMAND FAULTS            ,       ,          ,CE    /B  ,PWR    ,COM          ,N        ,US
   0          0              0      D    2.70E-02  2.60E-02  1.67E-02  4.05E-02    2.00
-----------------------------------------------------------------------------------------------------------------------------------------------------------------------
MECHANICAL          ,PUMP                     ,TURBINE DRIVEN               ,                                         ,FAILS TO RUN                                ,STANDBY
                          /       ,NUREG 1205/         ,DOES NOT OPERATE WITHOUT COMMAND FAULTS            ,       ,          ,B&W   /A  ,PWR    ,COM          ,N        ,US
   0          0              0      D    2.50E-02  2.41E-02  1.55E-02  3.75E-02    2.00
-----------------------------------------------------------------------------------------------------------------------------------------------------------------------
```

Figure 5.1 *Example CARP input sheet. From Science Applications International Corporation*

KEY TO TERMS ON CARP INPUT SHEET

CLASSIFICATION	=	Highest level taxonomy category, e.g., mechanical
COMPONENT TYPE	=	Equipment name, e.g., pump or valve
SUBTYPES 1 and 2	=	Subsequent taxonomy levels describing type of equipment, é.g., turbine driven pump
FAILURE MODE	=	Failure mode selection for aggregated data
OPERATING MODE	=	Normal mode of equipment operation, e.g., running or standby
SYSTEM ID/CODE	=	Plant system the equipment belongs to, e.g., feedwater
SOURCE/PAGE	=	Alphanumeric identifier of data source and page of source data was extracted from
SOURCE FAILURE MODE	=	Failure mode given for the data in the source
UNIT ID	=	ID number of the unit from which the data was taken; in most cases this is a generic identifier used to maintain agreed-upon anonymity of data
COMP ID	=	Alphanumeric ID number unique to a specific component
NSSS/CODE	=	For nuclear plants; refers to the nuclear steam supply vendor (e.g., GE for General Electric) and a code for the model/version of the plant
RXT	=	For nuclear plants; reactor type; BWR for Boiling Water reactor, PWR for pressurized water reactor
RXA	=	Reactor application, e.g., COM for commercial power generation
INDUSTRY	=	Code indicating industry the data was taken from, e.g., N for nuclear
GEOGRAPHIC	=	Country of data origin
#FAIL	=	Number of failures
DEMAND	=	Number of demands, for demand failure probabilities
EXP TIME	=	Exposure time, for time-related failure rates
UNIT	=	H for hourly failure rate, D for demand failure probability
MEAN, MEDIAN, 5TH 95TH, ERROR FACTOR	=	Parameters for the lognormally distributed data

sets and not necessarily the true behavior of a specific type of equipment in a particular application. The upper and lower bounds are not to be used to bracket the data conservatively, but rather to provide insight into the quality of the input data sets.

To incorporate both the location and spread of the distributions, CARP preserves the central tendency and upper bound of the distribution when generating a matching lognormal distribution. The lognormal distribution was chosen because of the general shape, popularity among data analysts, and ease of calculation. The results generated give a very good representation of the range of the data tails (5th and 95th percentiles). Since the resulting data will be used in an exponential model, the lognormal distribution was chosen to model the spread of the data—or the lack of homogeneity—of the data set. The exponential distribution is typically used to model the future behavior of the process equipment.

It should be noted that CARP generates an interim aggregated data set and then uses the central tendency and the upper bounds to match the data to a lognormal distribution. A final aggregated data set is then generated from this lognormal distribution. Often this results in a final aggregated lower bound which is lower than those of the input data sets or in the interim aggregated. Using such an aggregated lower would imply a lower failure rate than could be justified from the input data. The decision was made to select the maximum value of both the lower and upper bounds for failure rate data sheet entries.

The report writing function of CARP prints output data sheets such as that shown in Figure 5.2. These reports document the data resources used and decisions made to produce the data tables. The output data fields correspond to the input data fields listed at the beginning of Section 5.2 and shown in Figure 5.1. The mean values and associated error bounds of the aggregated data were used in the data tables presented in Section 5.5, except where higher levels of the CCPS Taxonomy are involved. For these latter cases, only error bounds were presented, since presenting a mean value for the large variety of data would have been misleading.

5.3 Data Table Presentation

Section 5.5 presents a data sheet for each cell in the taxonomy that contains failure rate data. Empty data cells are not presented. Filled data cells are listed by their CCPS Taxonomy number in Table 5.2 as an aid to the user. The CCPS data sheet format was developed from a number of sources including OREDA and IEEE Std. 500-1984. The format is presented in Figure 5.3, and its data elements are explained below:

Taxonomy number: The precise address of the data cell as defined by the classification scheme of the CCPS Taxonomy; each successive number indicates a successively lower level in the taxonomy.

Equipment description: Defines the equipment type that the data applies to; data resource equipment descriptions were used to match data to the descriptions at the taxonomy levels.

Operating mode: The operational service the equipment primarily experiences; expressed as alternating, running, or standby modes and reflected in the exposure hours or demands of that component.

Process severity: The indication of the degree of aggressiveness of the process medium on the hardware; expressed as categories 1 through 4, which correspond to Clean, General Industry, Moderately Severe, and Severe, respectively.

CARP -- DATA ANALYSIS DETAILED REPORT

Component Type Code: AV Component Name: AIR-OPERATED VALVE
Failure Mode Type Code: KR Failure Mode: SPURIOUS OPERATION

	D	MEAN	LOWER	MEDIAN	UPPER	P1	P2
Plant-specific							
Interim aggregated		3.02-06	9.17-09	1.03-06	1.11-05		
Aggregated generic	L	3.02-06	1.46-07	1.27-06	1.11-05	8.70+00	
Bayesian updated							
Final	L	3.02-06	1.46-07	1.27-06	1.11-05	8.70+00	

PLANT-SPECIFIC DATA
Units (N for demands, H for hours, etc.): H
Number of failures:
Exposure (time or number of demands):

BAYESIAN UPDATING
Bayesian updating performed: N

FINAL
Final basis (P,G,B): G
Final distribution type (L,G,B): L

KEY: L = Lognormal
P = Plant specific
G = Generic
B = Bayseian

AGGREGATION DETAILS
Aggregation method (T,A,G): T Weighting method (E,I,P,U,S): E

		D	MEAN	LOWER	MEDIAN	UPPER	P1	P2	QUALITY WEIGHT
1	IEEE-500	L	1.10-07				1.50+01		0.200
			1.10-07	1.89-09	2.84-08	4.25-07	1.50+01		
	Note: PAGE 1024 ; ALL MODES ; ASSUMED EF ; FAILURES ASSUMED DUE TO OPERATOR								
2	IPRD	L	2.20-06	5.90-07		5.70-06			0.200
			2.20-06	4.84-07	1.66-06	5.70-06	3.43+00		
	Note: NUREG/CR-3154 ; TABLE 9 ; DATA IS FOR PWRS								
3	OCONEE	L	5.15-06			2.45-05			0.200
			5.15-06	8.85-08	1.33-06	1.99-05	1.50+01		
	Note: PAGE 5-20 ; 1 FAILURE IN 1.94+05 HOURS								
4	OREDA-84	L	7.05-07			1.82-06			0.200
			7.05-07	1.57-07	5.34-07	1.82-06	3.41+00		
	Note: PAGE 97 ; 3 FAILURES IN 4253100 HOURS ; SIGNIFICANT LKG ; FIRE DELUGE VLV								
5	OREDA-84	L	6.92-06			1.58-05			0.200
			6.92-06	2.07-06	5.72-06	1.58-05	2.76+00		
	Note: PAGE 167 ; 4 FAILURES IN 578200 HOURS ; CONTROL VALVE								

Key to Abbreviations

D = Distribution
L = Lognormal
P1 = Parameter 1 = error factor
P2 = Placeholder for Parameter 2
T = Tolerance aggregation method
A, G = Modules for alternate aggregation methods
E = Equal weighting scheme
I, P, U, S = Codes available for alternate weighting scheme models

Figure 5.2 *Example CARP output sheet. From Science Applications International Corporation.*

TABLE 5.2
Index of Filled Data Cells

Code	Description
1.1.1	Motors-AC
1.1.1.1	Motors-AC-Induction
1.1.2	Motors-DC
1.2.1.1	Batteries-Lead Acid
1.2.1.2	Batteries-Nickel Cadmium
1.2.2	Battery Chargers
1.2.3.1	Circuit Breakers-AC
1.2.3.2	Circuit Breakers-DC
1.2.4	Inverters
1.2.6	Fuses
1.2.7.2	Relays-Protective
1.2.8.1	Transformers-Power
1.2.8.3	Transformers-Rectifier
1.3.1.1	Emergency Power Generators-Diesel Driven
2.1.1	Analyzers
2.1.3.1.2.3	Transmitters-Electronic-Level-Capacitance Probe
2.1.3.2.1	Transmitters-Pneumatic-Flow
2.1.3.2.1.1	Transmitters-Pneumatic-Flow-Differential Pressure
2.1.3.2.1.2	Transmitters-Pneumatic-Flow-Variable Area
2.1.3.2.2	Transmitters-Pneumatic-Level
2.1.3.2.2.1	Transmitters-Pneumatic-Level-Differential Pressure
2.1.3.2.2.3	Transmitters-Pneumatic-Level-Float
2.1.3.2.3	Transmitters-Pneumatic-Pressure
2.1.3.*.4	Transmitters-Temperature
2.1.3.*.5	Transmitters-Differential Pressure [Process Severity 2 & 3]
2.1.4.1.1	Switches-Electric-Flow
2.1.4.1.2	Switches-Electric-Level
2.1.4.1.3	Switches-Electric-Pressure
2.1.4.1.4	Switches-Electric-Temperature
2.1.4.1.7	Switches-Electric-Speed
2.1.4.2.1	Switches-Pneumatic-Flow
2.1.4.2.2	Switches-Pneumatic-Level
2.1.4.2.3	Switches-Pneumatic-Pressure
2.1.4.2.4	Switches-Pneumatic-Tempera ture
2.1.5	Flame Detectors
2.1.6.4.6	Indicators-Temperature-Radiation Pyrometer
2.1.8.1	Transducers-Current to Pneumatic
2.2.1	Controllers
2.2.1.1	Controllers-Electronic Panel board (Single Loop)
2.2.1.2	Controllers-Pneumatic Panelboard (Single Loop)
2.2.2	Annunciators
2.2.4	Recorders
2.2.5	Computational Modules-Pneumatic
3.1.2.2.1.1	Heat Transfer Devices-Non-Fired-InDirect Contact-Tubed-Baffled
3.2.1.1	Piping Systems-Metal-Straight Sections
3.2.1.2	Piping Systems-Metal-Fittings
3.2.1.4	Piping Systems-Metal-Connections
3.2.1.5.2	Piping Systems-Metal-Welds >1/2" to 4"
3.2.2.1	Piping Systems-Lined Pipe-Straight Sections
3.2.3.1	Piping Systems-Rigid Plastic-Straight Sections
3.2.5	Hoses
3.3.2	Rotating Equipment-Compressors
3.3.2.1	Rotating Equipment-Compressors-Electic Motor Driven
3.3.2.3	Rotating Equipment-Compressors-Turbine Driven
3.3.4	Rotating Equipment-Motor Driven Fans
3.3.7.2.1.1	Rotating Equipment-Pumps-Motor Driven-Pressure-Centrifugal (Alternating, Running, Standby)
3.3.7.3	Rotating Equipment-Pumps-Turbine Driven
3.4.2.2	Solids Handling-Conveyors-Screw
3.5.1.1	Valves-Operated-Stop Check
3.5.1.2	Valves-Non-Operated-Check
3.5.2	Valves-Manual
3.5.3.2	Valves-Operated-Motor
3.5.3.3	Valves-Operated-Pneumatic
3.5.3.4	Valves-Operated-Solenoid
3.6.1.1	Vessels-Atmospheric-Metallic
3.6.1.2	Vessels-Atmospheric-Non-Metallic
3.6.2.1	Vessels-Pressurized-Metallic
4.2.2	Protection Systems-Fire-Fire Detection
4.2.3.2	Protection Systems-Fire-Fire Suppression Systems-Water
4.2.3.3	Protection Systems-Fire-Fire Suppression Systems-Dry Powder
4.2.4.1	Protection Systems-Fire-Fire Water Pumps-Diesel
4.2.4.2	Protection Systems-Fire-Fire Water Pumps-Electric
4.3.3.1	Pressure-Safety Relief Valves-Pilot Operated
4.3.3.2	Pressure-Safety Relief Valves-Spring Loaded

DATA ON SELECTED PROCESS SYSTEMS AND EQUIPMENT	
Taxonomy No.	Equipment Description
Operating Mode	Process Severity

Population	Samples	Aggregated time in service (10^6 hrs)			No. of Demands	
		Calendar time	Operating time			
Failure mode	Failures (per 10^6 hrs)			Failures (per 10^3 demands)		
	Lower	Mean	Upper	Lower	Mean	Upper

Equipment Boundary
Data Reference No. (Table 5.1):

Figure 5.3 *Example data sheet for data cells*

Population: The total number of items of one particular type of equipment in service during the data window.

Sample: A failure rate data set for specific hardware in a given service.

Aggregated time in service **(10^6** hr): The calendar and/or operating time considered for the data denominator development, expressed in terms of 1 million hours (standby or running).

Number of demands: The number of actual or estimated challenges placed upon a component to perform its function within the data window; the demand-related failure probability denominator.

Failure mode: A symptom, condition, or fashion in which hardware fails. A mode might be identified as a loss of function; premature function (function without demand); and out of tolerance condition; or a simple physical characteristic such as a leak (incipient failure mode) observed during inspection.

Failures per: (10^6 hrs or 10^3 dmds) The lower bound, mean, and upper bound values of data expressed in terms of 1 million hours or one thousand demands.

Equipment boundary: Demarcation of the equipment showing components included and interfaces with excluded piping, electrical and instrumentation systems. Numbers shown on electrical equipment are taken from American Standard Device Function Numbers.

Comments: Remarks considered essential for the understanding and application of the data presented.

Data reference number: Refers to Table 5.1, which cites the data resources used to provide the failure rates presented in the data cell.

5.4 Use of the CCPS Generic Failure Rate Data Base

As explained in Section 3.3, failure rate data for a piece of equipment or system can be located by the taxonomy number for the equipment. The number can be found by using the CCPS Taxonomy, Appendix A, or the alphabetized hardware list in the Equipment Index, Appendix B. Table 5.2 shows whether the CCPS data base contains failure rate data for that numbered data cell or for an appropriate higher-level cell. Alternatively, the user may look directly for the desired taxonomy cell in the data tables.

When failure rate data are found in the data tables, the risk analyst must exercise good judgment in their use. The analyst may choose to use the data if the equipment description, process condition, and failure mode defined in the data cell are similar to the equipment being studied. More likely, the analyst will have to adjust the data to account for differences in equipment design, process conditions, properties of the chemicals being processed, severity of duty or quality of the facility maintenance regime, etc. Most of these factors, listed in Table 3.2, were not included as discrete levels in the CCPS Taxonomy but may heavily influence the data modification the analyst deems necessary. Little documented information exists to help the engineer adjust the data. The assistance of an expert may be required. Based on the discussion in Chapter 2, it is probably appropriate to apply adjustments only to the first significant number and associated exponential power for generic failure data.

5.5 CCPS Generic Data Tables

The pages in this section present tables of generic failure rate data compiled for process equipment and organized by the CCPS Taxonomy.

DATA ON SELECTED PROCESS SYSTEMS AND EQUIPMENT	
Taxonomy No. 1.1.1	Equipment Description MOTORS - AC
Operating Mode	Process Severity UNKNOWN

Population	Samples	Aggregated time in service (10^6 hrs)		No. of Demands
		Calendar time	Operating time	

Failure mode	Failures (per 10^6 hrs)			Failures (per 10^3 demands)		
	Lower	Mean	Upper	Lower	Mean	Upper
CATASTROPHIC a. Fails to Run Once Started b. Fails to Position Properly c. Fails to Start on Demand DEGRADED a. Shaft Fails to Run at Rated Speed b. Fails to Position Correctly c. "Hunts" for Correct Position	0.0222	15.2	46.7	0.00448	0.0247	0.0685

Equipment Boundary

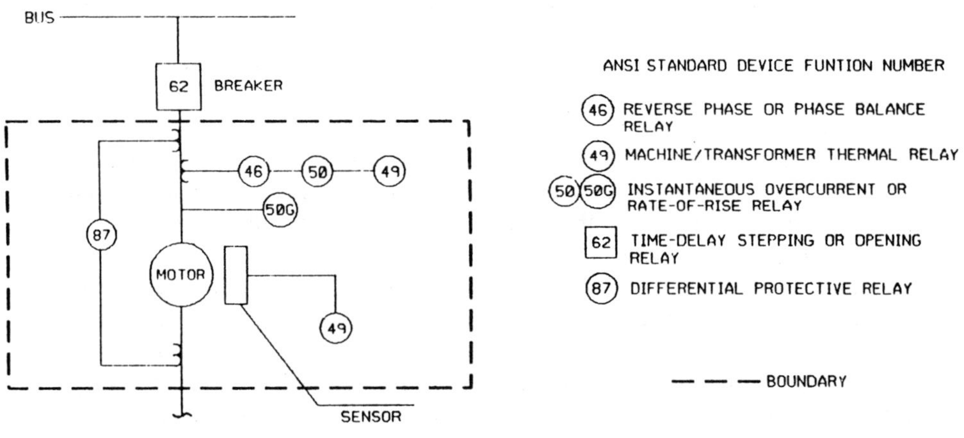

Data Reference No. (Table 5.1): 8.2, 8.5

DATA ON SELECTED PROCESS SYSTEMS AND EQUIPMENT

Taxonomy No.	1.1.1.1	Equipment Description	MOTORS - AC - INDUCTION
Operating Mode		Process Severity	UNKNOWN

Population	Samples	Aggregated time in service (10^6 hrs)		No. of Demands	
		Calendar time	Operating time		

Failure mode	Failures (per 10^6 hrs)			Failures (per 10^3 demands)		
	Lower	Mean	Upper	Lower	Mean	Upper
CATASTROPHIC a. Fails to Run Once Started b. Fails to Position Properly c. Fails to Start on Demand DEGRADED a. Shaft Fails to Run at Rated Speed b. Fails to Position Correctly c. "Hunts" for Correct Position	0.311	3.20	10.5	0.00448	0.0247	0.0685

Equipment Boundary

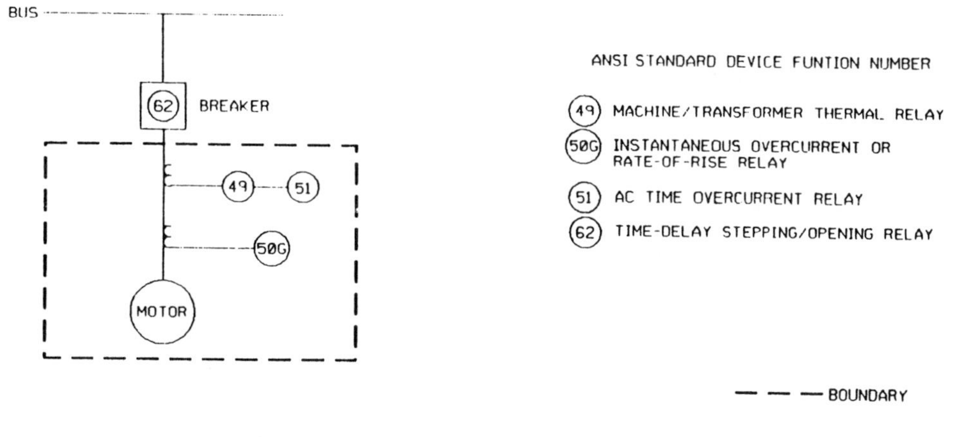

Data Reference No. (Table 5.1): 8.2, 8.5

DATA ON SELECTED PROCESS SYSTEMS AND EQUIPMENT	
Taxonomy No. 1.1.2	**Equipment Description** MOTORS - DC
Operating Mode	**Process Severity** UNKNOWN

Population	**Samples**	**Aggregated time in service (10^6 hrs)**		**No. of Demands**	
		Calendar time	**Operating time**		
Failure mode	**Failures (per 10^6 hrs)**			**Failures (per 10^3 demands)**		
	Lower	**Mean**	**Upper**	**Lower**	**Mean**	**Upper**
CATASTROPHIC	7.91	22.5	47.6			
a. Fails to Run Once Started						
b. Fails to Position Properly						
c. Fails to Start on Demand						
DEGRADED						
a. Shaft Fails to Run at Rated Speed						
b. Fails to Position Correctly						
c. "Hunts" for Correct Position						

Equipment Boundary

Data Reference No. (Table 5.1): 8.2, 8.5

DATA ON SELECTED PROCESS SYSTEMS AND EQUIPMENT

Taxonomy No. 1.2.1.1	Equipment Description BATTERIES - LEAD ACID
Operating Mode	Process Severity UNKNOWN

Population	Samples	Aggregated time in service (10^6 hrs)		No. of Demands
		Calendar time	Operating time	

Failure mode	Failures (per 10^6 hrs)			Failures (per 10^3 demands)		
	Lower	Mean	Upper	Lower	Mean	Upper
CATASTROPHIC						
a. No Output Given Challenge				5.77	13.2	24.8
b. No Output	0.379	2.25	6.42			
DEGRADED						
a. Low Output During Challenge						
b. No Output During Test						
c. Low Output During Test						
INCIPIENT						
a. Leakage						
b. Physical Evidence of Damage or Deterioration						

Equipment Boundary

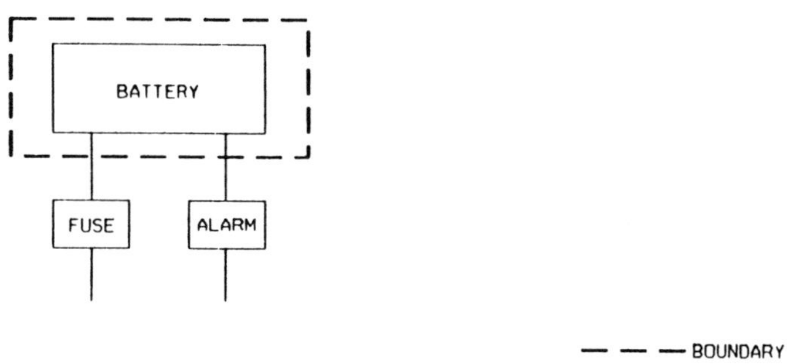

Data Reference No. (Table 5.1): 7, 8.1, 8.2, 8.3, 8.4, 8.15

DATA ON SELECTED PROCESS SYSTEMS AND EQUIPMENT

Taxonomy No. 1.2.1.2	Equipment Description BATTERIES - NICKEL CADMIUM
Operating Mode	Process Severity UNKNOWN

Population	Samples	Aggregated time in service (10^6 hrs)		No. of Demands
		Calendar time	Operating time	

Failure mode	Failures (per 10^6 hrs)			Failures (per 10^3 demands)		
	Lower	Mean	Upper	Lower	Mean	Upper
CATASTROPHIC a. No Output Given Challenge	0.220	0.251	0.285			
DEGRADED a. Low Output During Challenge b. No Output During Test c. Low Output During Test						
INCIPIENT a. Leakage b. Physical Evidence of Damage or Deterioration						

Equipment Boundary

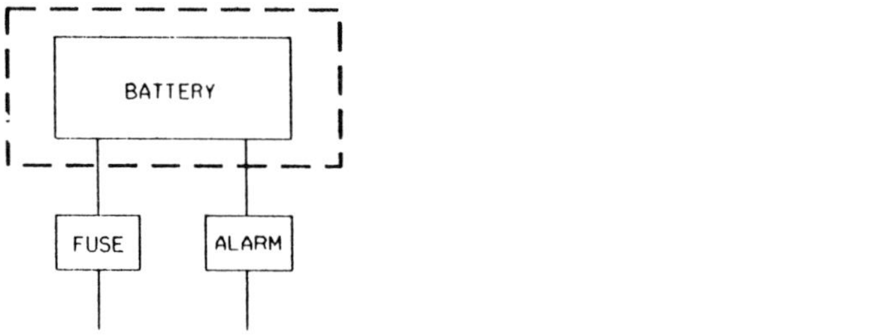

Data Reference No. (Table 5.1): 8.5

DATA ON SELECTED PROCESS SYSTEMS AND EQUIPMENT	
Taxonomy No. 1.2.2	**Equipment Description** BATTERY CHARGERS
Operating Mode	**Process Severity** UNKNOWN

Population	**Samples**	**Aggregated time in service (10^6 hrs)**		**No. of Demands**		
		Calendar time	**Operating time**			
Failure mode	**Failures (per 10^6 hrs)**			**Failures (per 10^3 demands)**		
	Lower	**Mean**	**Upper**	**Lower**	**Mean**	**Upper**
CATASTROPHIC						
a. No Electrical Output	0.284	7.60	28.5			
DEGRADED						
a. Low Output						
b. High Output						
c. Erratic Output						
INCIPIENT						
a. Overheated						

Equipment Boundary

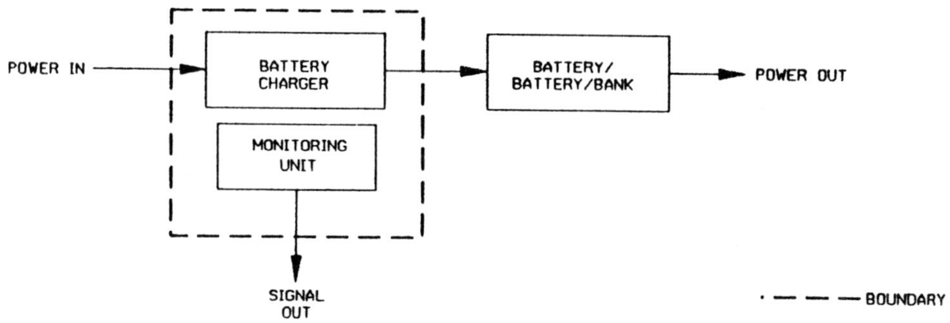

Data Reference No. (Table 5.1): 8.1, 8.2, 8.3, 8.4, 8.11, 8.12

DATA ON SELECTED PROCESS SYSTEMS AND EQUIPMENT	
Taxonomy No. 1.2.3.1	**Equipment Description** CIRCUIT BREAKERS - AC
Operating Mode	**Process Severity** UNKNOWN

Population	**Samples**	**Aggregated time in service (10^6 hrs)**		**No. of Demands**	
		Calendar time	**Operating time**		

Failure mode	**Failures (per 10^6 hrs)**			**Failures (per 10^3 demands)**		
	Lower	**Mean**	**Upper**	**Lower**	**Mean**	**Upper**
CATASTROPHIC a. Spurious Operation b. Failed to Open on Demand c. Failed to Close on Demand	 0.162	 1.75	 5.79	0.203	1.16	3.24
DEGRADED						
INCIPIENT a. Contaminated						

Equipment Boundary

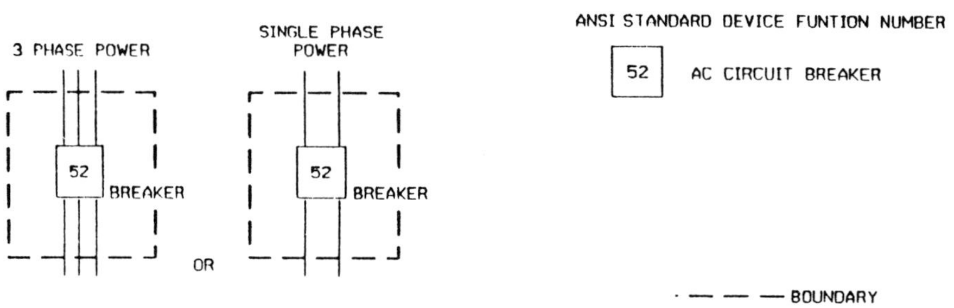

Data Reference No. (Table 5.1): 7, 8.1, 8.2, 8.3, 8.4, 8.6, 8.11, 8.12, 8.14, 8.15

DATA ON SELECTED PROCESS SYSTEMS AND EQUIPMENT

Taxonomy No.	1.2.3.2	Equipment Description	CIRCUIT BREAKERS - DC
Operating Mode		Process Severity	UNKNOWN

Population	Samples	Aggregated time in service (10^6 hrs)		No. of Demands
		Calendar time	Operating time	

Failure mode	Failures (per 10^6 hrs)			Failures (per 10^3 demands)		
	Lower	Mean	Upper	Lower	Mean	Upper
CATASTROPHIC						
a. Spurious Operation	0.0348	3.80	14.4			
b. Failed to Open on Demand				0.0927	0.883	2.85
c. Failed to Close on Demand						
DEGRADED						
INCIPIENT						
a. Contaminated						

Equipment Boundary

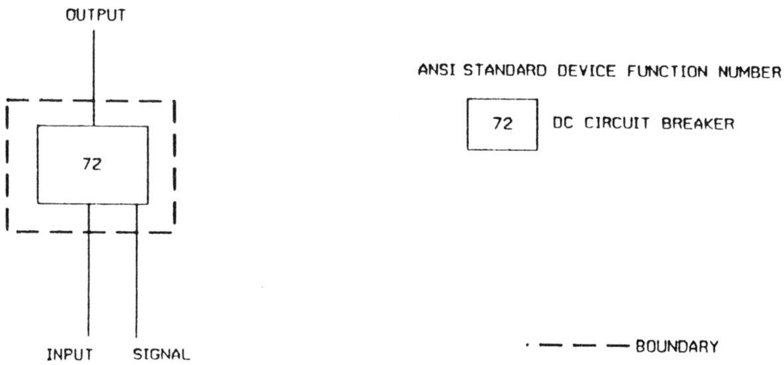

Data Reference No. (Table 5.1): 8.2, 8.3, 8.15

DATA ON SELECTED PROCESS SYSTEMS AND EQUIPMENT

Taxonomy No.	1.2.4	Equipment Description	INVERTERS
Operating Mode		Process Severity	UNKNOWN

Population	Samples	Aggregated time in service (10^6 hrs)		No. of Demands	
		Calendar time	Operating time		

Failure mode	Failures (per 10^6 hrs)			Failures (per 10^3 demands)		
	Lower	Mean	Upper	Lower	Mean	Upper
CATASTROPHIC						
a. No Output	1.04	28.7	116.0			
b. Fails to Transfer						
DEGRADED						
a. Output Frequency Outside Spec.						
b. Output Voltage Outside Spec.						
c. Output Current Outside Spec.						
INCIPIENT						
a. Overheating						
b. Faulty Indication						
c. Dirt/dust Contamination						

Equipment Boundary

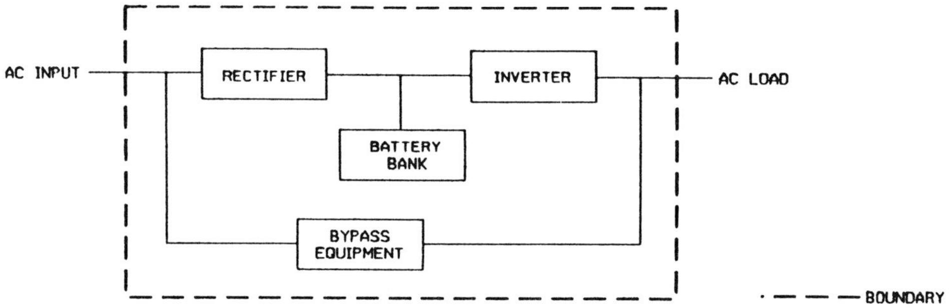

Data Reference No. (Table 5.1): 8.1, 8.3, 8.4, 8.10, 8.12, 8.14

DATA ON SELECTED PROCESS SYSTEMS AND EQUIPMENT			
Taxonomy No.	1.2.6	**Equipment Description**	FUSES
Operating Mode		**Process Severity**	UNKNOWN

Population	**Samples**	**Aggregated time in service (10^6 hrs)**		**No. of Demands**	
		Calendar time	**Operating time**		

Failure mode	**Failures (per 10^6 hrs)**			**Failures (per 10^3 demands)**		
	Lower	**Mean**	**Upper**	**Lower**	**Mean**	**Upper**
CATASTROPHIC a. Premature Open	0.0265	0.634	2.36			

Equipment Boundary

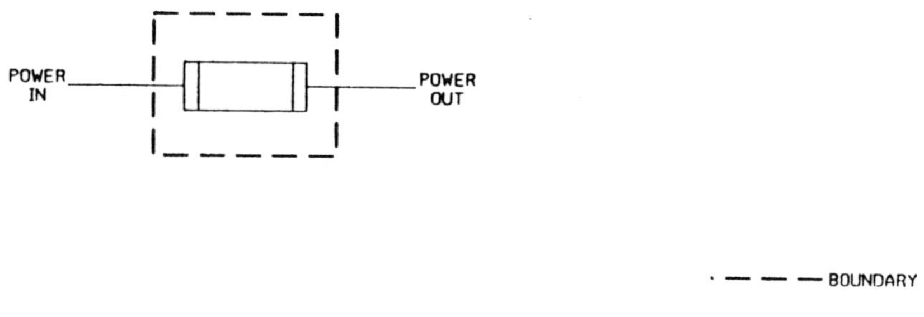

Data Reference No. (Table 5.1): 8.2, 8.3, 8.5, 8.15

DATA ON SELECTED PROCESS SYSTEMS AND EQUIPMENT	
Taxonomy No. 1.2.7.2	**Equipment Description** RELAYS - PROTECTIVE
Operating Mode	**Process Severity** UNKNOWN

Population	**Samples**	**Aggregated time in service (10^6 hrs)**		**No. of Demands**
		Calendar time	**Operating time**	

Failure mode	**Failures (per 10^6 hrs)**			**Failures (per 10^3 demands)**		
	Lower	**Mean**	**Upper**	**Lower**	**Mean**	**Upper**
CATASTROPHIC	1.79	1.91	2.04			
a. Fails to Close on Demand						
b. Fails to Open on Demand						
c. Spurious Operation	0.00104	0.06	0.232			
d. Delayed Change of State	0.0015	0.00288	0.00486			
e .Premature Change of State	0.00387	0.00598	0.0087			

Equipment Boundary

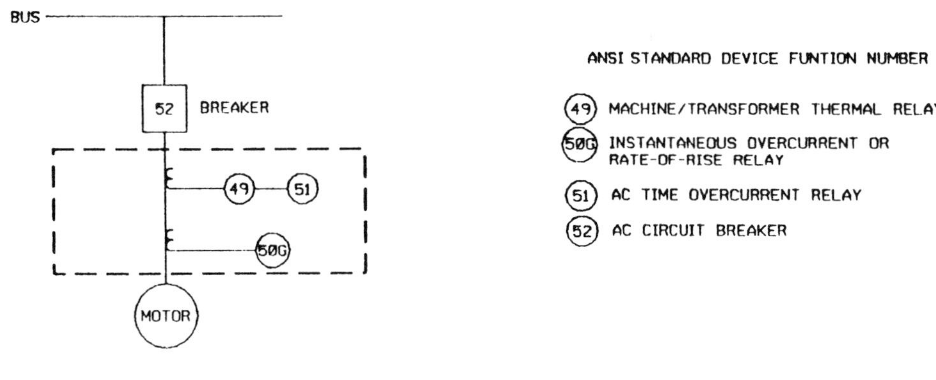

Data Reference No. (Table 5.1): 8.2, 8.5, 8.7

DATA ON SELECTED PROCESS SYSTEMS AND EQUIPMENT

Taxonomy No. 1.2.8.1	Equipment Description TRANSFORMERS - POWER
Operating Mode	Process Severity UNKNOWN

Population	Samples	Aggregated time in service (10^6 hrs)		No. of Demands		
		Calendar time	Operating time			

Failure mode	Failures (per 10^6 hrs)			Failures (per 10^3 demands)		
	Lower	Mean	Upper	Lower	Mean	Upper
CATASTROPHIC						
a. No Output	0.125	2.53	9.26			
DEGRADED						
a. Incorrect Output						
INCIPIENT						
a. Insulation Integrity						

Equipment Boundary

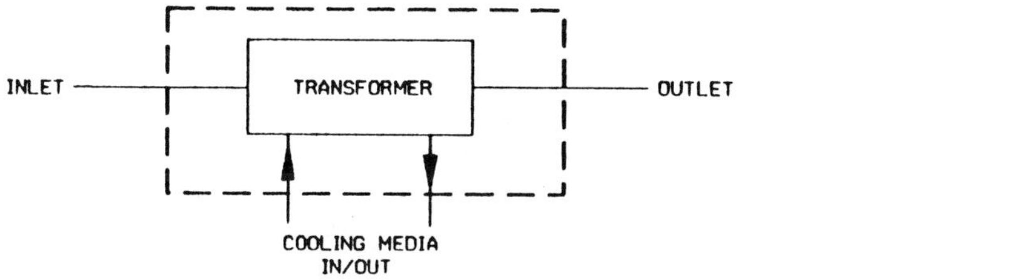

Data Reference No. (Table 5.1): 7, 8.4, 8.14

DATA ON SELECTED PROCESS SYSTEMS AND EQUIPMENT			
Taxonomy No.	1.2.8.3	**Equipment Description**	TRANSFORMERS - RECTIFIER
Operating Mode		**Process Severity**	UNKNOWN

Population	**Samples**	**Aggregated time in service (10^6 hrs)**		**No. of Demands**
		Calendar time	**Operating time**	

Failure mode	**Failures (per 10^6 hrs)**			**Failures (per 10^3 demands)**		
	Lower	**Mean**	**Upper**	**Lower**	**Mean**	**Upper**
CATASTROPHIC						
a.No Output	0.357	1.07	2.31			
DEGRADED						
a. Incorrect Output						
INCIPIENT						
a. Insulation Integrity						

Equipment Boundary

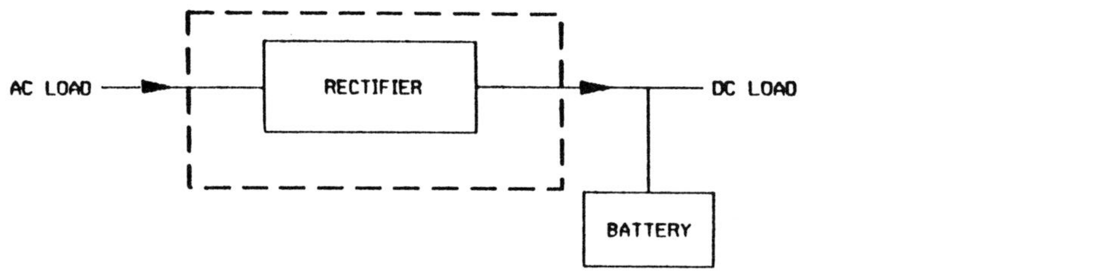

Data Reference No. (Table 5.1): 7, 8.2, 8.4

DATA ON SELECTED PROCESS SYSTEMS AND EQUIPMENT	
Taxonomy No. 1.3.1.1	**Equipment Description** EMERGENCY POWER GENERATORS-DIESEL DRIVEN
Operating Mode STANDBY	**Process Severity** UNKNOWN

Population	**Samples**	**Aggregated time in service (10^6 hrs)**		**No. of Demands**
		Calendar time	**Operating time**	

Failure mode	**Failures (per 10^6 hrs)**			**Failures (per 10^3 demands)**		
	Lower	**Mean**	**Upper**	**Lower**	**Mean**	**Upper**
CATASTROPHIC a. Fails to Start b. Fails to Run c. Fails to Supply Load in Time	172.0	2250.0	7710.0	2.22	17.6	54.5
DEGRADED a. Delayed Start after Multiple Attempts b. Fails to Maintain Voltage/ Frequency Specification						
INCIPIENT a. Improper Cooling/Heating b. RPM Hunting c. Faulty Indication d. Vibration e. Improper Lubrication						

Equipment Boundary

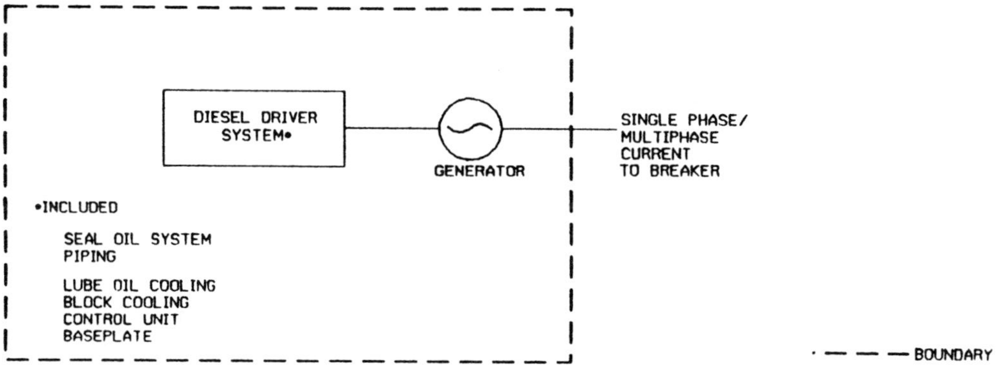

Comment: Based on nuclear power plant emergency diesel service.

Data Reference No. (Table 5.1): 8, 8.1, 8.2, 8.3, 8.9, 8.10, 8.11, 8.13, 8.14, 8.15

DATA ON SELECTED PROCESS SYSTEMS AND EQUIPMENT	
Taxonomy No. 2.1.1	**Equipment Description** ANALYZERS
Operating Mode	**Process Severity** UNKNOWN

Population	**Samples**	**Aggregated time in service (10^6 hrs)**		**No. of Demands**	
		Calendar time	**Operating time**		

Failure mode	**Failures (per 10^6 hrs)**			**Failures (per 10^3 demands)**		
	Lower	**Mean**	**Upper**	**Lower**	**Mean**	**Upper**
CATASTROPHIC	6.87		4800.0			
a. Maximum Output						
b. No Output						
c. No Change of Output with Change of Input						
DEGRADED						
a. Erratic Output						
b. High Output						
c. Low Output						
INCIPIENT						

Equipment Boundary

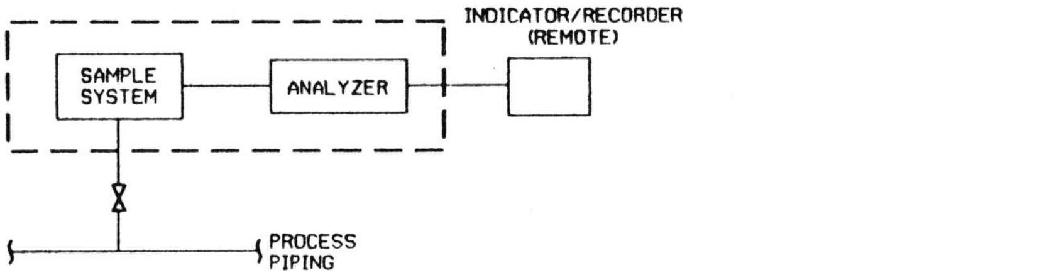

Data Reference No. (Table 5.1): 4

DATA ON SELECTED PROCESS SYSTEMS AND EQUIPMENT

Taxonomy No. 2.1.3.1.2.3	**Equipment Description** TRANSMITTERS-ELECTRONIC LEVEL- CAPACITANCE PROBE
Operating Mode	**Process Severity** UNKNOWN

Population	**Samples**	**Aggregated time in service (10^6 hrs)**		**No. of Demands**
		Calendar time	**Operating time**	

Failure mode	**Failures (per 10^6 hrs)**			**Failures (per 10^3 demands)**		
	Lower	**Mean**	**Upper**	**Lower**	**Mean**	**Upper**
CATASTROPHIC a. Maximum Output b. No Output c. No Change of Output with Change of Input	0.436	25.1	97.1			
DEGRADED a. Erratic Output b. High Output c. Low Output						
INCIPIENT						

Equipment Boundary

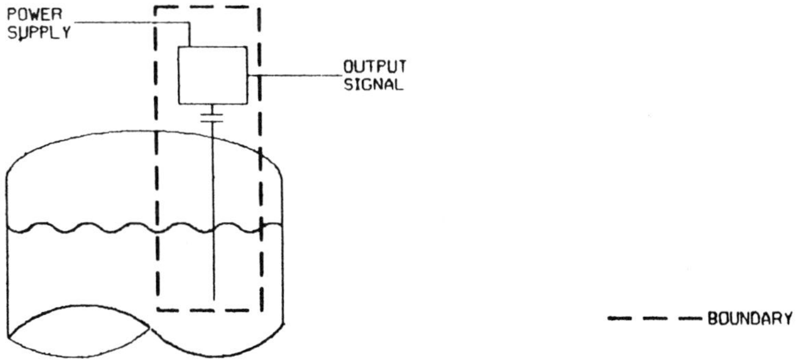

Data Reference No. (Table 5.1): 4

DATA ON SELECTED PROCESS SYSTEMS AND EQUIPMENT

Taxonomy No. 2.1.3.2.1	Equipment Description TRANSMITTERS - PNEUMATIC-FLOW
Operating Mode	Process Severity UNKNOWN

Population	Samples	Aggregated time in service (10^6 hrs)		No. of Demands		
		Calendar time	Operating time			
Failure mode		Failures (per 10^6 hrs)			Failures (per 10^3 demands)	
	Lower	Mean	Upper	Lower	Mean	Upper
CATASTROPHIC a. Maximum Output b. No Output c. No Change of Output with Change of Input DEGRADED a. Erratic Output b. High Output c. Low Output	1.93	109.0	439.0			

Equipment Boundary

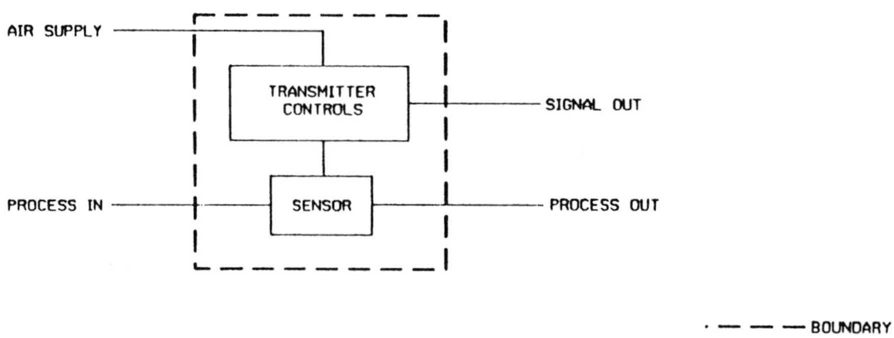

Data Reference No. (Table 5.1): 4, 7

DATA ON SELECTED PROCESS SYSTEMS AND EQUIPMENT

Taxonomy No. 2.1.3.2.1.1	Equipment Description TRANSMITTERS - PNEUMATIC FLOW-DIFFERENTIAL PRESSURE
Operating Mode	Process Severity UNKNOWN

Population	Samples	Aggregated time in service (10^6 hrs) Calendar time	Operating time	No. of Demands

Failure mode	Failures (per 10^6 hrs) Lower	Mean	Upper	Failures (per 10^3 demands) Lower	Mean	Upper
CATASTROPHIC (includes control faults)	2.03	118.0	486.0			
a. Maximum Output						
b. No Output						
c. No Change of Output with Change of Input						
DEGRADED						
a. Erratic Output						
b. High Output						
c. Low Output						
INCIPIENT						

Equipment Boundary

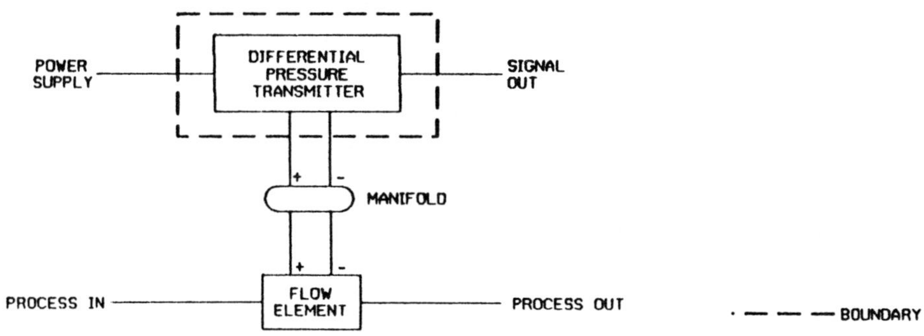

Data Reference No. (Table 5.1): 4, 7

DATA ON SELECTED PROCESS SYSTEMS AND EQUIPMENT

Taxonomy No. 2.1.3.2.1.2	Equipment Description TRANSMITTERS-PNEUMATIC-FLOW-VARIABLE AREA
Operating Mode	Process Severity UNKNOWN

Population	Samples	Aggregated time in service (10^6 hrs)		No. of Demands
		Calendar time	Operating time	

Failure mode	Failures (per 10^6 hrs)			Failures (per 10^3 demands)		
	Lower	Mean	Upper	Lower	Mean	Upper
CATASTROPHIC a. Maximum Output b. No Output c. No Change of Output with Change of Input	1.59	96.3	373.0			
DEGRADED a. Erratic Output b. High Output c. Low Output						
INCIPIENT						

Equipment Boundary

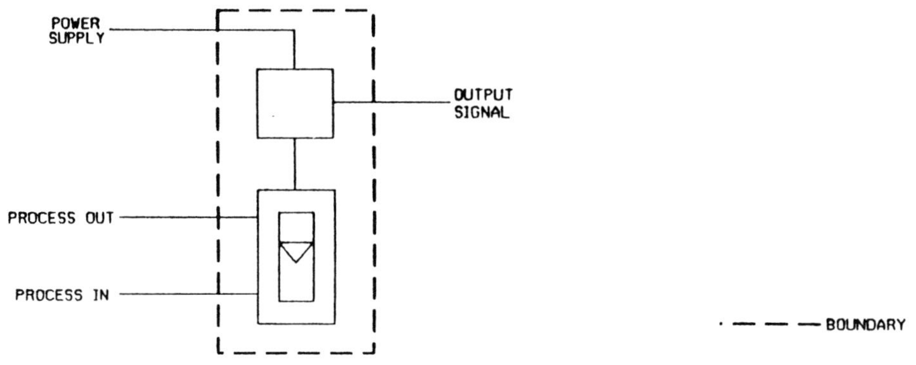

Data Reference No. (Table 5.1): 4

DATA ON SELECTED PROCESS SYSTEMS AND EQUIPMENT

Taxonomy No. 2.1.3.2.2	Equipment Description TRANSMITTERS-PNEUMATIC-LEVEL
Operating Mode	Process Severity UNKNOWN

Population	Samples	Aggregated time in service (10^6 hrs)		No. of Demands	
		Calendar time	Operating time		

Failure mode	Failures (per 10^6 hrs)			Failures (per 10^3 demands)		
	Lower	Mean	Upper	Lower	Mean	Upper
CATASTROPHIC a. Maximum Output b. No Output c. No Change of Output with Change of Input	2.32	141.0	573.0			
DEGRADED a. Erratic Output b. High Output c. Low Output						
INCIPIENT						

Equipment Boundary

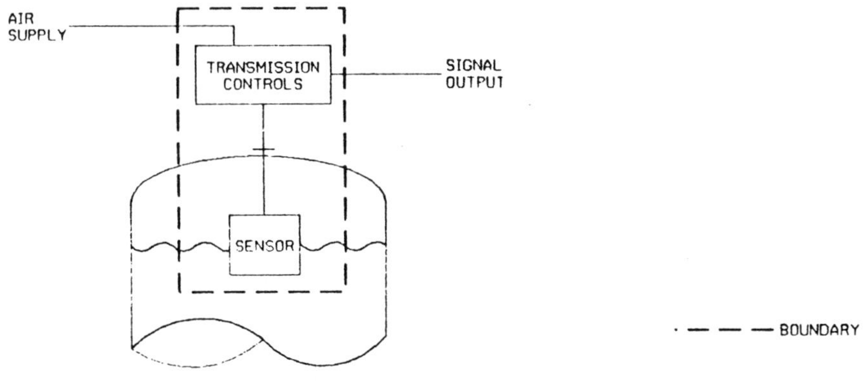

Data Reference No. (Table 5.1): 4,7

DATA ON SELECTED PROCESS SYSTEMS AND EQUIPMENT	
Taxonomy No. 2.1.3.2.2.1	**Equipment Description** TRANSMITTERS-PNEUMATIC-LEVEL-DIFFERENTIAL PRESSURE
Operating Mode	**Process Severity** UNKNOWN

Population	**Samples**	**Aggregated time in service (10^6 hrs)**		**No. of Demands**	
		Calendar time	**Operating time**		

Failure mode	**Failures (per 10^6 hrs)**			**Failures (per 10^3 demands)**		
	Lower	**Mean**	**Upper**	**Lower**	**Mean**	**Upper**
CATASTROPHIC	2.18	99.4	417.0			
a. Maximum Output						
b. No Output						
c. No Change of Output with Change of Input						
DEGRADED						
a. Erratic Output						
b. High Output						
c. Low Output						
INCIPIENT						

Equipment Boundary

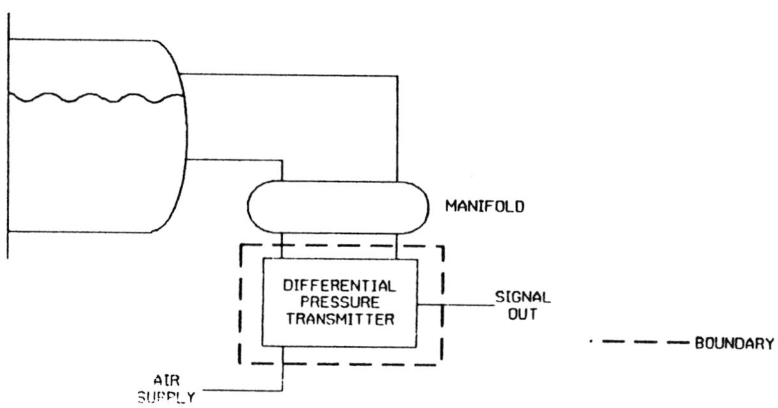

Data Reference No. (Table 5.1): 4,7

DATA ON SELECTED PROCESS SYSTEMS AND EQUIPMENT	
Taxonomy No. 2.1.3.2.2.3	**Equipment Description** TRANSMITTERS-PNEUMATIC-LEVEL-FLOAT
Operating Mode	**Process Severity** UNKNOWN

Population	**Samples**	**Aggregated time in service (10^6 hrs)**			**No. of Demands**	
		Calendar time		**Operating time**		
Failure mode		**Failures (per 10^6 hrs)**			**Failures (per 10^3 demands)**		
		Lower	**Mean**	**Upper**	**Lower**	**Mean**	**Upper**
CATASTROPHIC a. Maximum Output b. No Output c. No Change of Output with Change of Input		3.25	187.0	723.0			
DEGRADED a. Erratic Output b. High Output c. Low Output							
INCIPIENT							

Equipment Boundary

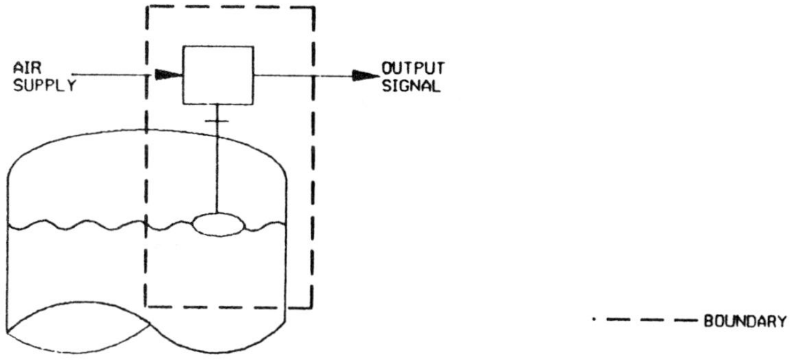

Data Reference No. (Table 5.1): 4

DATA ON SELECTED PROCESS SYSTEMS AND EQUIPMENT

Taxonomy No. 2.1.3.2.3	Equipment Description TRANSMITTERS-PNEUMATIC-PRESSURE
Operating Mode	Process Severity UNKNOWN

Population	Samples	Aggregated time in service (10^6 hrs)		No. of Demands
		Calendar time	Operating time	

Failure mode	Failures (per 10^6 hrs)			Failures (per 10^3 demands)		
	Lower	Mean	Upper	Lower	Mean	Upper
CATASTROPHIC a. Maximum Output b. No Output c. No Change of Output with Change of Input	0.159	91.3	381.0			
DEGRADED a. Erratic Output b. High Output c. Low Output						
INCIPIENT						

Equipment Boundary

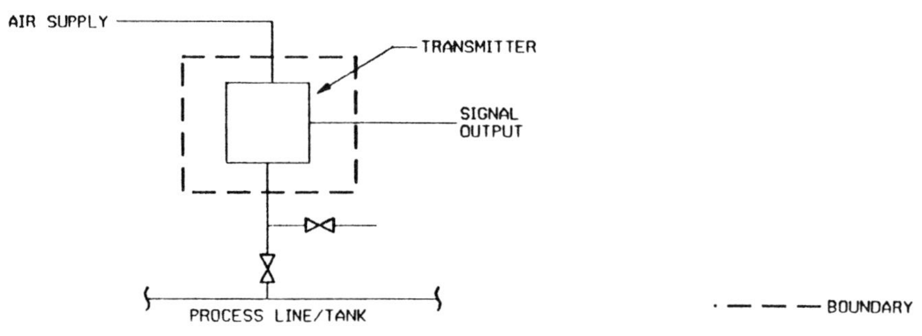

Data Reference No. (Table 5.1): 4,7

DATA ON SELECTED PROCESS SYSTEMS AND EQUIPMENT

Taxonomy No. 2.1.3.*.4	Equipment Description TRANSMITTERS- TEMPERATURE
Operating Mode	Process Severity UNKNOWN

Population	Samples	Aggregated time in service (10^6 hrs)		No. of Demands
		Calendar time	Operating time	

Failure mode	Failures (per 10^6 hrs)			Failures (per 10^3 demands)		
	Lower	Mean	Upper	Lower	Mean	Upper
CATASTROPHIC	1.68	97.0	375.0			
a. Maximum Output						
b. No Output						
c. No Change of Output with Change of Input						
DEGRADED						
a. Erratic Output						
b. High Output						
c. Low Output						
INCIPIENT						

Equipment Boundary

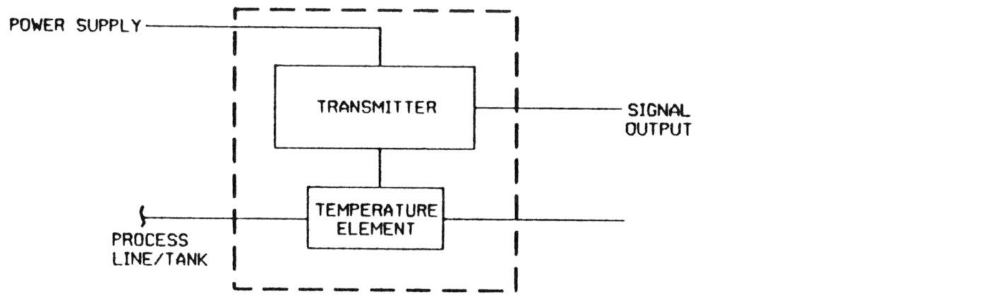

COMMENTS: Sources do not specify between pneumatic and electronic transmitters

Data Reference No. (Table 5.1): 4

DATA ON SELECTED PROCESS SYSTEMS AND EQUIPMENT

Taxonomy No. 2.1.3.*.5	Equipment Description TRANSMITTERS - DIFFERENTIAL PRESSURE
Operating Mode	Process Severity 2

Population	Samples	Aggregated time in service (10^6 hrs)		No. of Demands	
		Calendar time	Operating time		

Failure mode	Failures (per 10^6 hrs)			Failures (per 10^3 demands)		
	Lower	Mean	Upper	Lower	Mean	Upper
CATASTROPHIC a. Maximum Output b. No Output c. No Change of Output with Change of Input DEGRADED a. Erratic Output b. High Output c. Low Output INCIPIENT	1.01	65.6	254.0			

Equipment Boundary

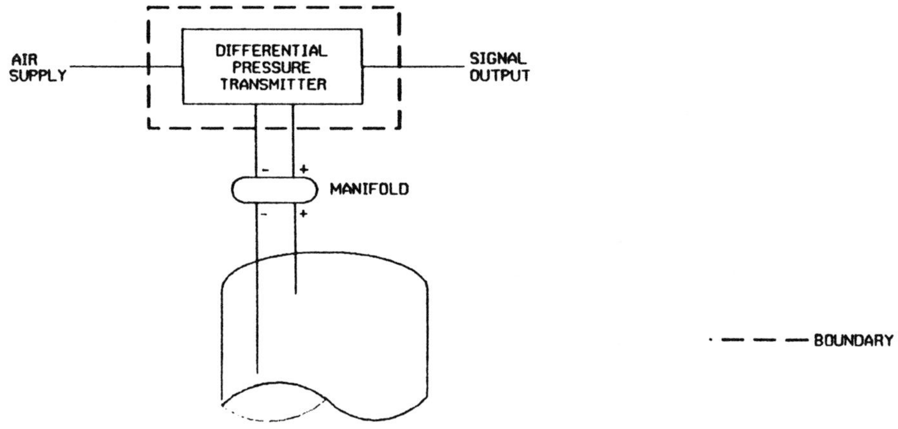

COMMENTS: Source does not specify between pneumatic and electronic transmitters

Data Reference No. (Table 5.1): 4

DATA ON SELECTED PROCESS SYSTEMS AND EQUIPMENT	
Taxonomy No. 2.1.3.*.5	Equipment Description TRANSMITTERS - DIFFERENTIAL PRESSURE
Operating Mode	Process Severity 3

Population	Samples	Aggregated time in service (10^6 hrs)		No. of Demands		
		Calendar time	Operating time			
Failure mode		**Failures (per 10^6 hrs)**			**Failures (per 10^3 demands)**	

Failure mode	Lower	Mean	Upper	Lower	Mean	Upper
CATASTROPHIC	3.79	218.0	843.0			
a. Maximum Output						
b. No Output						
c. No Change of Output with Change of Input						
DEGRADED						
a. Erratic Output						
b. High Output						
c. Low Output						
INCIPIENT						

Equipment Boundary

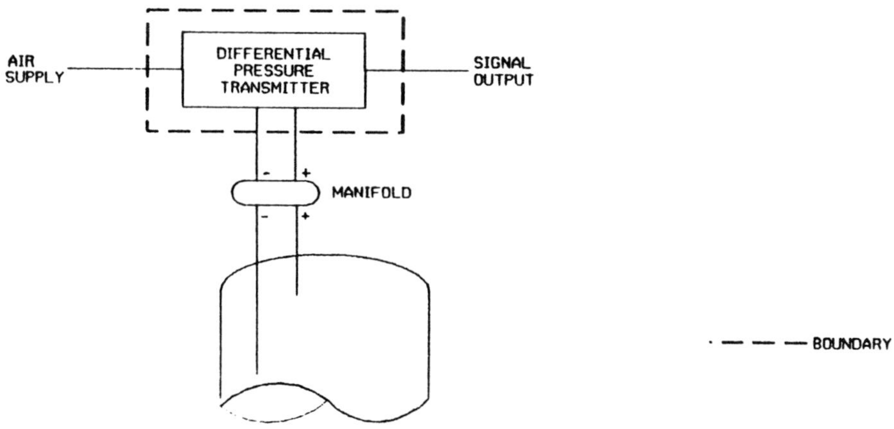

COMMENTS: Source does not specify between pneumatic and electronic transmitters

Data Reference No. (Table 5.1): 4

DATA ON SELECTED PROCESS SYSTEMS AND EQUIPMENT

Taxonomy No.	2.1.4.1.1	Equipment Description	SWITCHES -ELECTRIC - FLOW
Operating Mode		Process Severity	3

Population	Samples	Aggregated time in service (10^6 hrs) Calendar time	Operating time	No. of Demands

Failure mode	Failures (per 10^6 hrs) Lower	Mean	Upper	Failures (per 10^3 demands) Lower	Mean	Upper
CATASTROPHIC	0.917	26.8	101.0			
a. Functioned Without Signal	0.378	0.86	1.61			
b. Failed to Function when Signaled	0.165	4.20	15.7			
DEGRADED						
a. Functioned at Improper Signal Level						
b. Intermittent Operation						
INCIPIENT						
a. In-service Problems						

Equipment Boundary

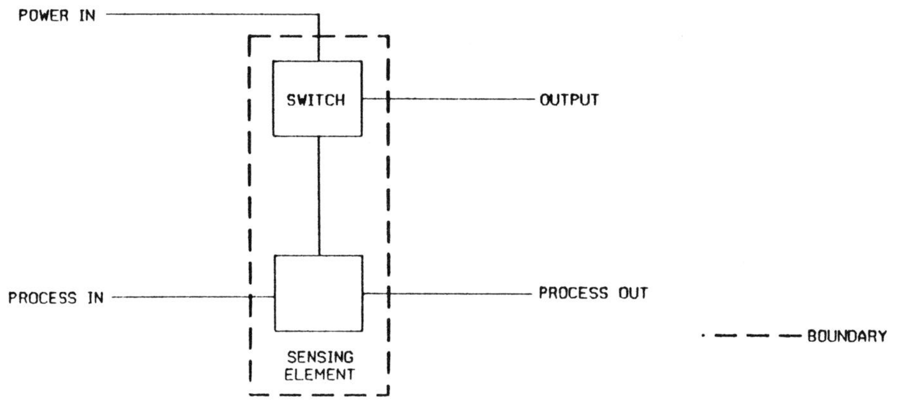

Comment: Process Severity 3 applies only to Catastrophic rate.

Data Reference No. (Table 5.1): 4, 8, 8.2, 8.4

DATA ON SELECTED PROCESS SYSTEMS AND EQUIPMENT

Taxonomy No. 2.1.4.1.2	Equipment Description SWITCHES - ELECTRIC - LEVEL
Operating Mode	Process Severity UNKNOWN

Population	Samples	Aggregated time in service (10^6 hrs)		No. of Demands
		Calendar time	Operating time	

Failure mode	Failures (per 10^6 hrs)			Failures (per 10^3 demands)		
	Lower	Mean	Upper	Lower	Mean	Upper
CATASTROPHIC	0.737	1.74	3.33			
a. Functioned without Signal	0.152	0.925	2.65			
b. Failed to Function when Signaled	0.0702	0.17	0.33			
DEGRADED	0.375	1.70	4.40			
a. Functioned at Improper Signal Level						
b. Intermittent Operation						
INCIPIENT						
a. In-service Problems						

Equipment Boundary

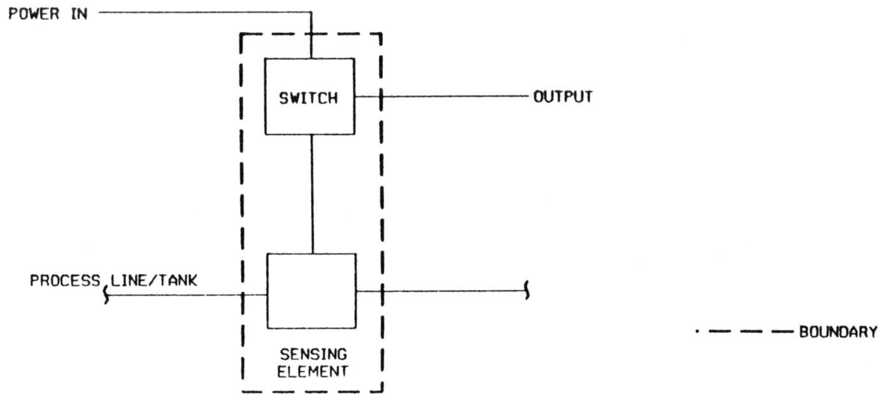

Data Reference No. (Table 5.1): 8.2, 8.4

DATA ON SELECTED PROCESS SYSTEMS AND EQUIPMENT	
Taxonomy No. 2.1.4.1.3	Equipment Description SWITCHES - ELECTRIC - PRESSURE
Operating Mode	Process Severity 3

Population	Samples	Aggregated time in service (10^6 hrs)		No. of Demands
		Calendar time	Operating time	

Failure mode	Failures (per 10^6 hrs)			Failures (per 10^3 demands)		
	Lower	Mean	Upper	Lower	Mean	Upper
CATASTROPHIC	0.525	49.6	192.0			
a. Functioned without Signal	0.00122	0.07	0.271			
b. Failed to Function when Signaled	0.00809	0.40	1.55			
DEGRADED						
a. Functioned at Improper Signal Level						
b. Intermittent Operation						
INCIPIENT						
a. In-service Problems						

Equipment Boundary

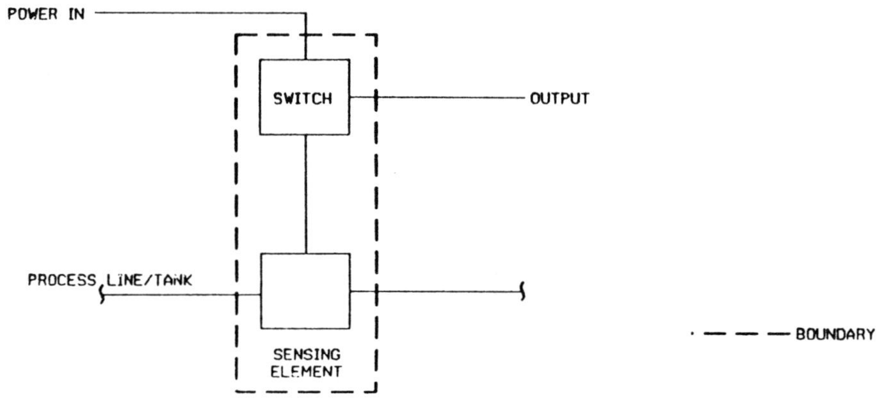

Comment: Process Severity 3 applies only to Catastrophic rate.

Data Reference No. (Table 5.1): 4, 8.2

DATA ON SELECTED PROCESS SYSTEMS AND EQUIPMENT			
Taxonomy No. 2.1.4.1.4		**Equipment Description** SWITCHES - ELECTRIC - TEMPERATURE	
Operating Mode		**Process Severity** UNKNOWN	

Population	Samples	Aggregated time in service (10^6 hrs)		No. of Demands		
		Calendar time	Operating time			
Failure mode	Failures (per 10^6 hrs)			Failures (per 10^3 demands)		
	Lower	Mean	Upper	Lower	Mean	Upper
CATASTROPHIC	0.102	2.28	10.1			
a. Functioned without Signal	0.107	1.16	4.69			
b. Failed to Function when Signaled	0.0971	3.40	13.4			
DEGRADED						
a. Functioned at Improper Signal Level						
b. Intermittent Operation						
INCIPIENT						
a. In-service Problems						

Equipment Boundary

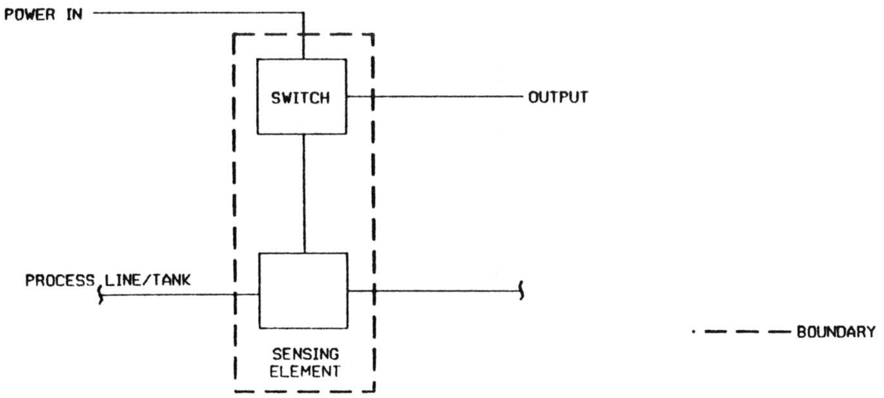

Data Reference No. (Table 5.1): 8.2, 8.4

DATA ON SELECTED PROCESS SYSTEMS AND EQUIPMENT

Taxonomy No.	2.1.4.1.7	Equipment Description	SWITCHES - ELECTRIC - SPEED
Operating Mode		Process Severity	UNKNOWN

Population	Samples	Aggregated time in service (10^6 hrs) Calendar time	Operating time	No. of Demands

Failure mode	Failures (per 10^6 hrs) Lower	Mean	Upper	Failures (per 10^3 demands) Lower	Mean	Upper
CATASTROPHIC	0.285	0.48	0.742			
a. Functioned without Signal	0.02	0.15	0.455			
b. Failed to Function when Signaled	0.541	0.57	0.60			
DEGRADED						
a. Functioned at Improper Signal Level						
b. Intermittent Operation						
INCIPIENT						
a. In-service Problems						

Equipment Boundary

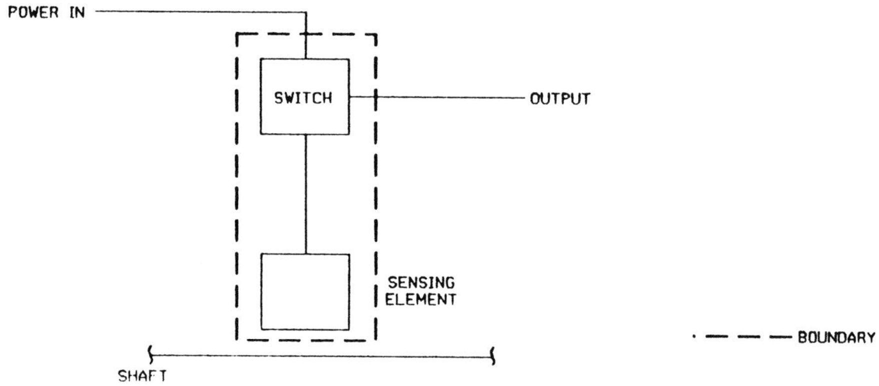

Data Reference No. (Table 5.1): 8.2

DATA ON SELECTED PROCESS SYSTEMS AND EQUIPMENT	
Taxonomy No. 2.1.4.2.1	Equipment Description SWITCHES - PNEUMATIC - FLOW
Operating Mode	Process Severity UNKNOWN

Population	Samples	Aggregated time in service (10^6 hrs)		No. of Demands		
		Calendar time	Operating time			
Failure mode		**Failures (per 10^6 hrs)**			**Failures (per 10^3 demands)**		
	Lower	Mean	Upper	Lower	Mean	Upper
CATASTROPHIC	1.64	4.0	7.8			
a. Functioned without Signal b. Failed to Function when Signaled	0.027	0.66	2.46			
DEGRADED a. Functioned at Improper Signal Level b. Intermittent Operation	0.15	1.30	4.10			
INCIPIENT a. In-service Problems						

Equipment Boundary

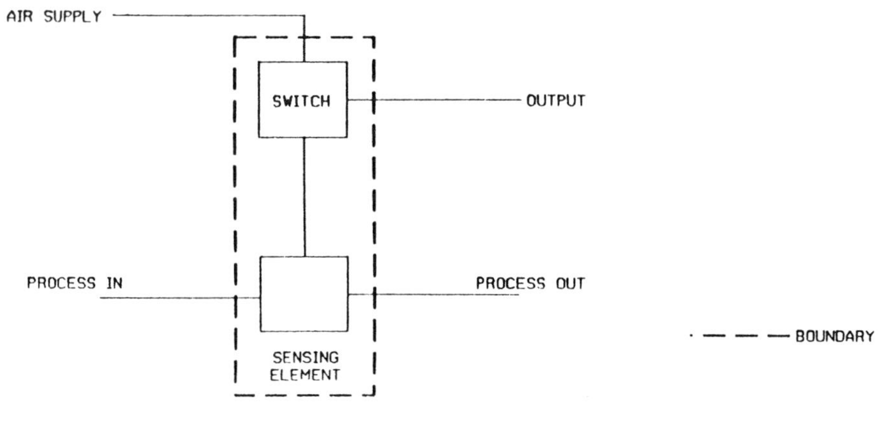

Data Reference No. (Table 5.1): 8.4

DATA ON SELECTED PROCESS SYSTEMS AND EQUIPMENT

Taxonomy No. 2.1.4.2.2	Equipment Description SWITCHES - PNEUMATIC - LEVEL
Operating Mode	Process Severity UNKNOWN

Population	Samples	Aggregated time in service (10^6 hrs)		No. of Demands
		Calendar time	Operating time	

Failure mode	Failures (per 10^6 hrs)			Failures (per 10^3 demands)		
	Lower	Mean	Upper	Lower	Mean	Upper
CATASTROPHIC a. Functioned without Signal	0.0972	0.62	1.80			
b.Failed to Function when Signaled	0.0484	0.54	1.80			
DEGRADED a. Functioned at Improper Signal Level	0.375	1.70	4.40			
b. Intermittent Operation						
INCIPIENT a. In-service Problems						

Equipment Boundary

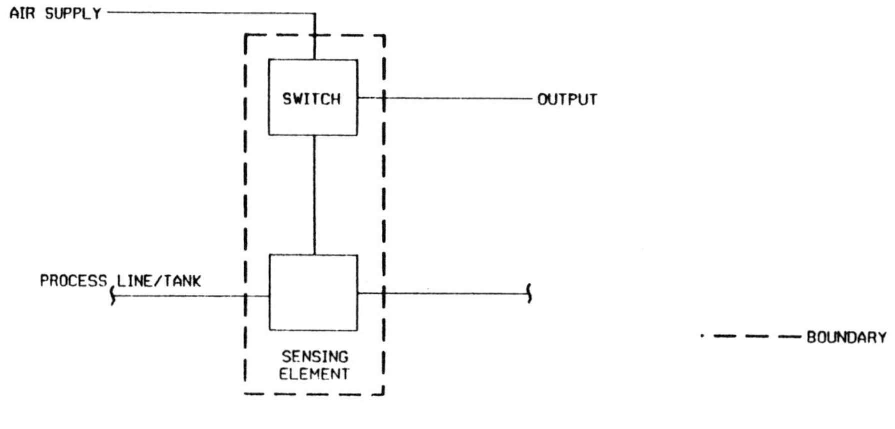

Data Reference No. (Table 5.1): 8.4

DATA ON SELECTED PROCESS SYSTEMS AND EQUIPMENT	
Taxonomy No. 2.1.4.2.3	**Equipment Description** SWITCHES -PNEUMATIC - PRESSURE
Operating Mode	**Process Severity** UNKNOWN

Population	Samples	Aggregated time in service (10^6 hrs) Calendar time	Operating time	No. of Demands

Failure mode	Failures (per 10^6 hrs) Lower	Mean	Upper	Failures (per 10^3 demands) Lower	Mean	Upper
CATASTROPHIC	2.18	5.20	10.0			
a. Functioned without Signal	0.0515	0.47	1.50			
b. Failed to Function when Signaled	0.11	1.70	6.0			
DEGRADED	11.8	18.0	26.0			
a. Functioned at Improper Signal Level						
b. Intermittent Operation						
INCIPIENT						
a. In-service Problems						

Equipment Boundary

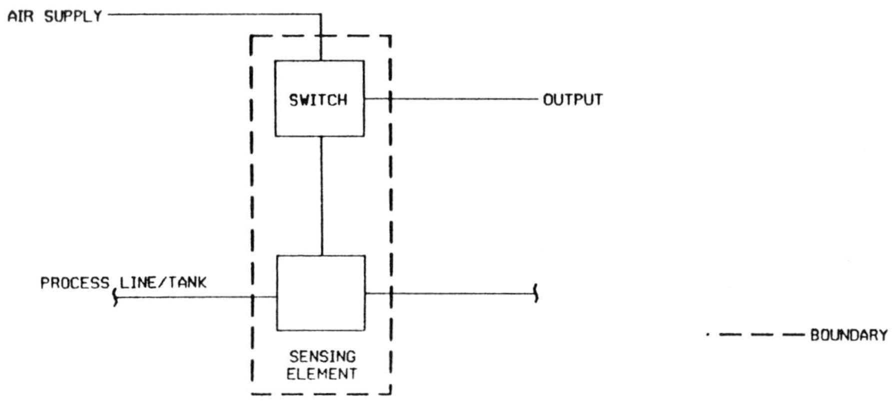

Data Reference No. (Table 5.1): 8.4

DATA ON SELECTED PROCESS SYSTEMS AND EQUIPMENT	
Taxonomy No. 2.1.4.2.4	**Equipment Description** SWITCHES - PNEUMATIC - TEMPERATURE
Operating Mode	**Process Severity** UNKNOWN

Population	**Samples**	**Aggregated time in service (10^6 hrs)**		**No. of Demands**	
		Calendar time	**Operating time**		

Failure mode	**Failures (per 10^6 hrs)**			**Failures (per 10^3 demands)**		
	Lower	**Mean**	**Upper**	**Lower**	**Mean**	**Upper**
CATASTROPHIC a. Functioned without Signal b. Failed to Function when Signaled	1.09	5.00	13.0			
DEGRADED a. Functioned at Improper Signal Level b. Intermittent Operation	0.61	3.00	8.0			
INCIPIENT a. In-service Problems						

Equipment Boundary

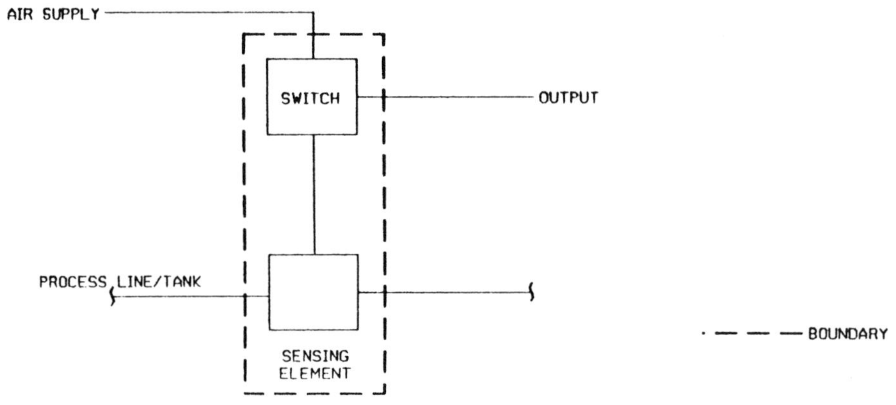

Data Reference No. (Table 5.1): 8.4

DATA ON SELECTED PROCESS SYSTEMS AND EQUIPMENT

Taxonomy No. 2.1.5	Equipment Description FLAME DETECTORS
Operating Mode	Process Severity UNKNOWN

Population	Samples	Aggregated time in service (10^6 hrs)		No. of Demands
		Calendar time	Operating time	

Failure mode	Failures (per 10^6 hrs)			Failures (per 10^3 demands)		
	Lower	Mean	Upper	Lower	Mean	Upper
CATASTROPHIC a. Functioned without Signal b. Failed to Function when Signaled	0.053	432.0	1760.0			
DEGRADED a. Functioned at Improper Signal Level b. Intermittent Operation						
INCIPIENT a. In-service Problems						

Equipment Boundary

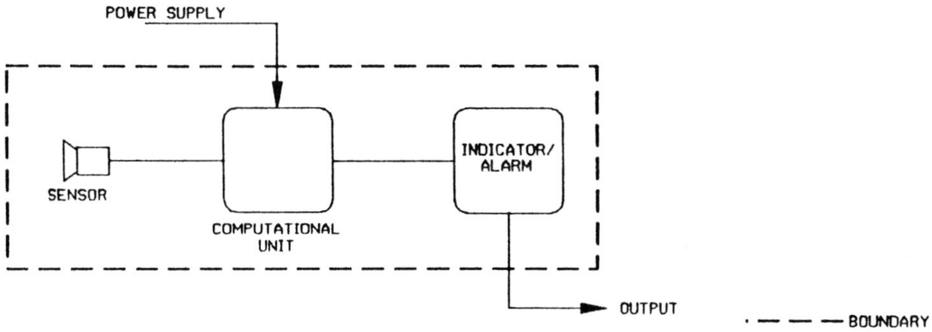

Data Reference No. (Table 5.1): 1, 4

DATA ON SELECTED PROCESS SYSTEMS AND EQUIPMENT

Taxonomy No. 2.1.6.4.6	Equipment Description INDICATORS - TEMPERATURE-RADIATION PYROMETER
Operating Mode	Process Severity UNKNOWN

Population	Samples	Aggregated time in service (10^6 hrs)		No. of Demands		
		Calendar time	Operating time			
Failure mode		Failures (per 10^6 hrs)			Failures (per 10^3 demands)		
		Lower	Mean	Upper	Lower	Mean	Upper
CATASTROPHIC a. Maximum Output b. No Output c. No Change of Output with Change of Input DEGRADED a. Erratic Output b. High Output c. Low Output INCIPIENT		4.31	248.0	959.0			

Equipment Boundary

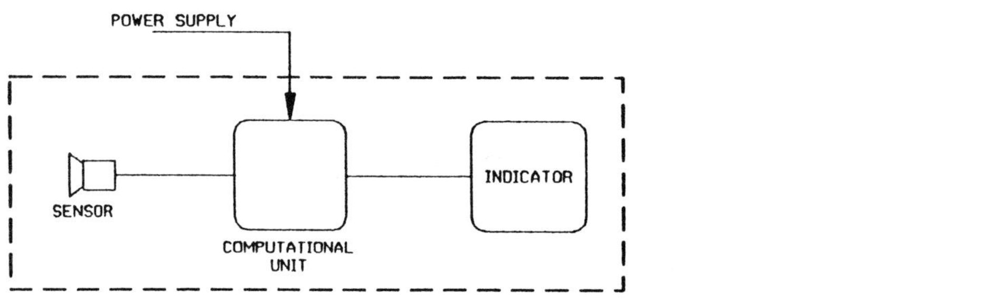

Data Reference No. (Table 5.1): 4

DATA ON SELECTED PROCESS SYSTEMS AND EQUIPMENT

Taxonomy No. 2.1.8.1	Equipment Description TRANSDUCERS - CURRENT TO PNEUMATIC
Operating Mode	Process Severity UNKNOWN

Population	Samples	Aggregated time in service (10^6 hrs) Calendar time	Operating time	No. of Demands	

Failure mode	Failures (per 10^6 hrs) Lower	Mean	Upper	Failures (per 10^3 demands) Lower	Mean	Upper
CATASTROPHIC a. Maximum Output b. No Output c. No Change of Output with Change of Input	1.08	62.8	243.0			
DEGRADED a. Erratic Output b. High Output c. Low Output						
INCIPIENT						

Equipment Boundary

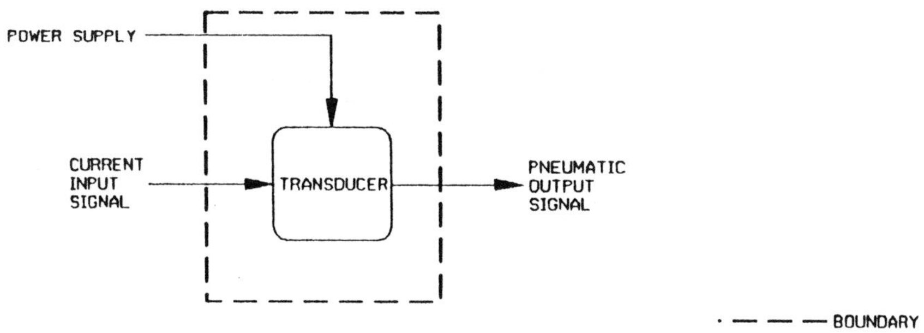

Data Reference No. (Table 5.1): 4

DATA ON SELECTED PROCESS SYSTEMS AND EQUIPMENT

Taxonomy No. 2.2.1	Equipment Description CONTROLLERS
Operating Mode	Process Severity UNKNOWN

Population	Samples	Aggregated time in service (10^6 hrs)		No. of Demands
		Calendar time	Operating time	

Failure mode	Failures (per 10^6 hrs)			Failures (per 10^3 demands)		
	Lower	Mean	Upper	Lower	Mean	Upper
CATASTROPHIC a.Maximum Output b. No Output c. No Change of Output with Change of Input DEGRADED a. Erratic Output b. High Output c. Low Output INCIPIENT	2.43	68.8	260.0			

Equipment Boundary

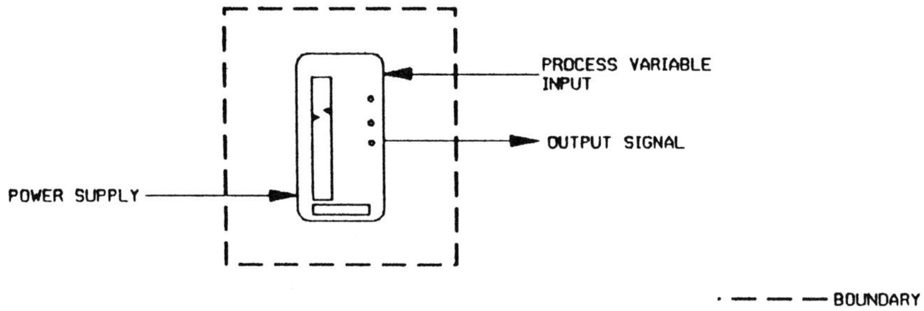

Data Reference No. (Table 5.1): 4

DATA ON SELECTED PROCESS SYSTEMS AND EQUIPMENT

Taxonomy No. 2.2.1.1	Equipment Description CONTROLLERS - ELECTRONIC PANELBOARD (SINGLE LOOP)
Operating Mode	**Process Severity** UNKNOWN

Population	Samples	Aggregated time in service (10^6 hrs)		No. of Demands
		Calendar time	Operating time	

Failure mode	Failures (per 10^6 hrs)			Failures (per 10^3 demands)		
	Lower	Mean	Upper	Lower	Mean	Upper
CATASTROPHIC	3.56	205.0	793.0			
a.Maximum Output						
b. No Output						
c. No Change of Output with Change of Input						
DEGRADED						
a. Erratic Output						
b. High Output						
c. Low Output						
INCIPIENT						

Equipment Boundary

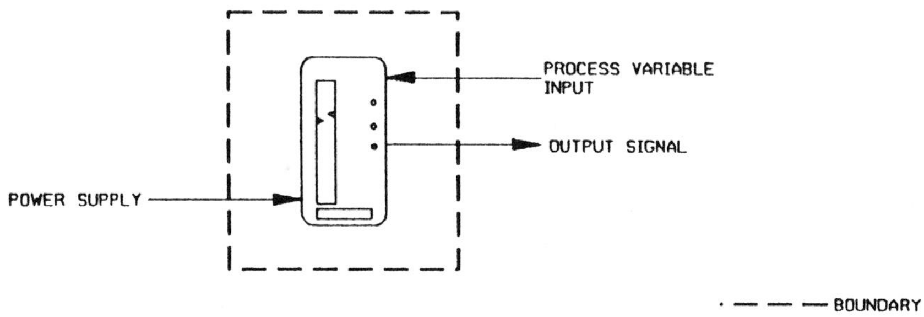

Data Reference No. (Table 5.1): 4

DATA ON SELECTED PROCESS SYSTEMS AND EQUIPMENT

Taxonomy No. 2.2.1.2	**Equipment Description** CONTROLLERS - PNEUMATIC PANELBOARD (SINGLE LOOP)
Operating Mode	**Process Severity** UNKNOWN

Population	**Samples**	**Aggregated time in service (10^6 hrs)**		**No. of Demands**
		Calendar time	**Operating time**	

Failure mode	**Failures (per 10^6 hrs)**			**Failures (per 10^3 demands)**		
	Lower	**Mean**	**Upper**	**Lower**	**Mean**	**Upper**
CATASTROPHIC a. Maximum Output b. No Output c. No Change of Output with Change of Input DEGRADED a. Erratic Output b. High Output c. Low Output	0.754	43.4	168.0			

Equipment Boundary

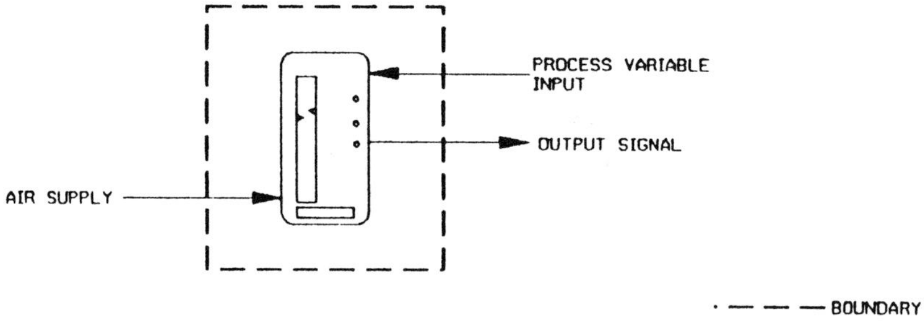

Data Reference No. (Table 5.1): 4

DATA ON SELECTED PROCESS SYSTEMS AND EQUIPMENT

Taxonomy No. 2.2.2	Equipment Description ANNUNCIATORS
Operating Mode	Process Severity UNKNOWN

Population	Samples	Aggregated time in service (10^6 hrs) Calendar time	Operating time	No. of Demands

Failure mode	Failures (per 10^6 hrs) Lower	Mean	Upper	Failures (per 10^3 demands) Lower	Mean	Upper
CATASTROPHIC	0.0272	0.77	2.90			
a. Operates Spuriously						
b. Fails to Operate on Demand						
DEGRADED						
a. Operates at Low Intensity						
INCIPIENT						
a. Arcing						

Equipment Boundary

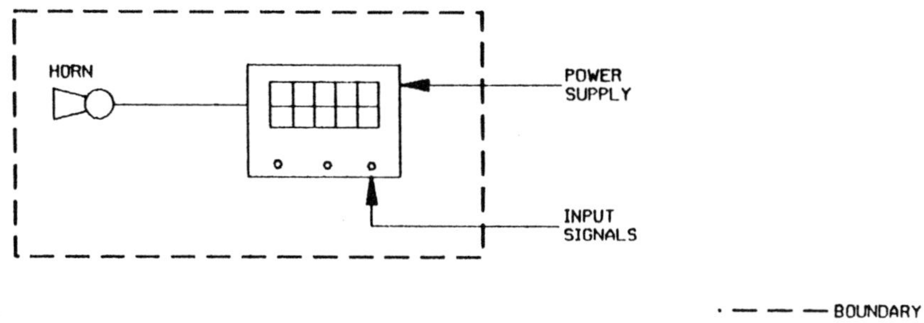

Data Reference No. (Table 5.1): 8.2

DATA ON SELECTED PROCESS SYSTEMS AND EQUIPMENT

Taxonomy No. 2.2.4	Equipment Description RECORDERS
Operating Mode	Process Severity UNKNOWN

Population	Samples	Aggregated time in service (10^6 hrs) Calendar time	Operating time	No. of Demands	

Failure mode	Failures (per 10^6 hrs) Lower	Mean	Upper	Failures (per 10^3 demands) Lower	Mean	Upper
CATASTROPHIC a. Maximum Output b. No Output c. No Change of Output with Change of Input	0.436	25.1	97.1			
DEGRADED a. Erratic Output b. High Output c. Low Output						
INCIPIENT						

Equipment Boundary

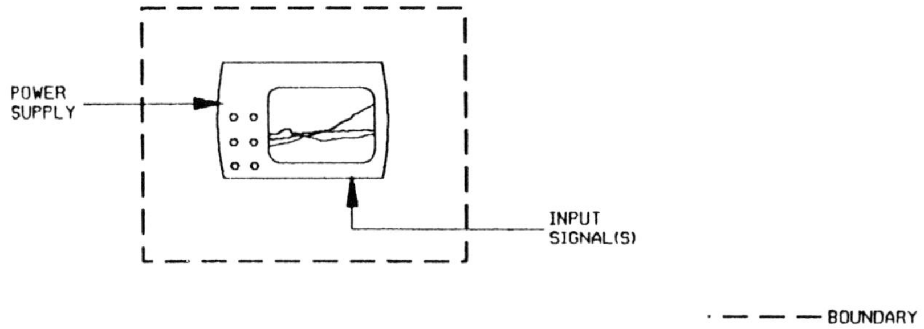

Data Reference No. (Table 5.1): 4

DATA ON SELECTED PROCESS SYSTEMS AND EQUIPMENT	
Taxonomy No. 2.2.5	Equipment Description COMPUTATIONAL MODULES-PNEUMATIC
Operating Mode	Process Severity UNKNOWN

Population	Samples	Aggregated time in service (10^6 hrs)		No. of Demands		
		Calendar time	Operating time			
Failure mode	Failures (per 10^6 hrs)			Failures (per 10^3 demands)		
	Lower	Mean	Upper	Lower	Mean	Upper
CATASTROPHIC a. Maximum Output b. No Output c. No Change of Output with Change of Input	0.337	19.4	75.0			
DEGRADED a. Erratic Output b. High Output c. Low Output						
INCIPIENT						

Equipment Boundary

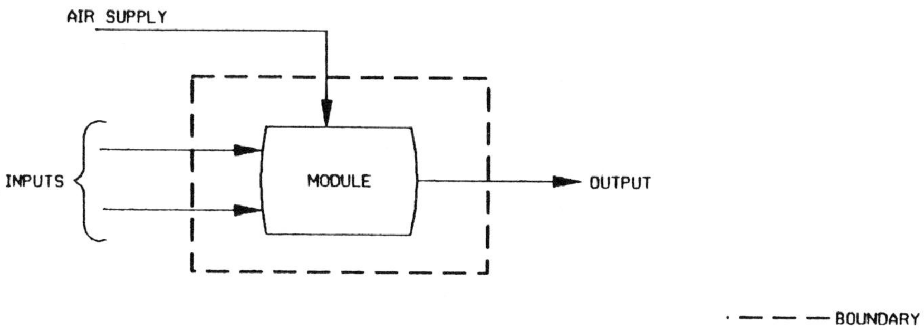

Data Reference No. (Table 5.1): 4

DATA ON SELECTED PROCESS SYSTEMS AND EQUIPMENT

Taxonomy No.	3.1.2.2.1.1	Equipment Description	HEAT TRANSFER DEVICES - NON-FIRED INDIRECT CONTACT-TUBED-BAFFLED
Operating Mode		Process Severity	UNKNOWN

Population	Samples	Aggregated time in service (10^6 hrs) Calendar time	Operating time	No. of Demands		

Failure mode	Failures (per 10^6 hrs) Lower	Mean	Upper	Failures (per 10^3 demands) Lower	Mean	Upper
CATASTROPHIC	0.0173	31.1	131.0			
a. Leakage 0 - 1/4"						
b. Leakage >1/4"	1.62	25.8	91.3			
c. Rupture						
DEGRADED						
INCIPIENT						
a. Wall Thinning						
b. Embrittlement						
c. Cracked or Flawed						

Equipment Boundary

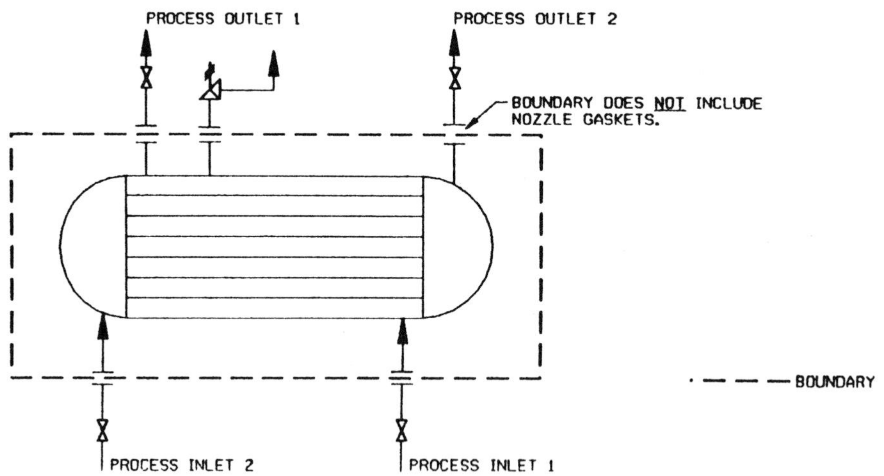

Data Reference No. (Table 5.1): 8.4, 8.13, 8.3

DATA ON SELECTED PROCESS SYSTEMS AND EQUIPMENT			
Taxonomy No. 3.2.1.1		Equipment Description	PIPING SYSTEMS -METAL - STRAIGHT SECTIONS
Operating Mode		Process Severity	UNKNOWN

Population	Samples	Aggregated time in service (10^6 hrs)		No. of Demands
		Calendar time	Operating time	

Failure mode	Failures (per 10^6 hrs)			Failures (per 10^3 demands)		
	Lower	Mean	Upper	Lower	Mean	Upper
CATASTROPHIC	0.000465	0.0268	0.104			
a. 0 - 10% Flow Area						
b. >10% Flow Area						
c. Rupture						
d. Plugged						
DEGRADED						
a. Restricted Flow						
INCIPIENT						
a. Wall Thinning						
b. Embrittlement						
c. Cracked or Flawed						
d. Erratic Flow						

Equipment Boundary

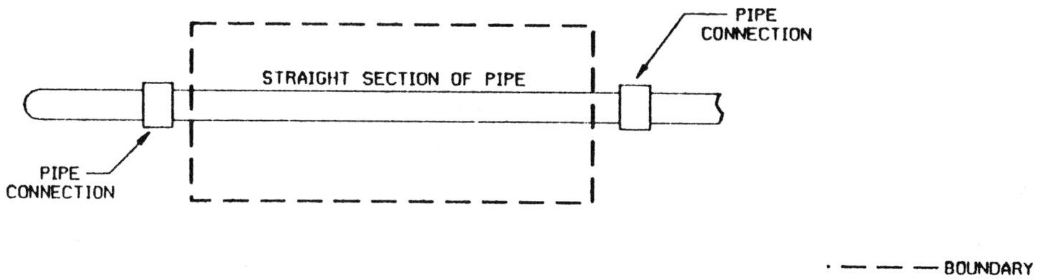

COMMENTS: DATA IS IN MILE-HOURS

Data Reference No. (Table 5.1): 9

DATA ON SELECTED PROCESS SYSTEMS AND EQUIPMENT	
Taxonomy No. 3.2.1.4	**Equipment Description** PIPING SYSTEMS- METAL-CONNECTIONS
Operating Mode	**Process Severity** UNKNOWN

Population	**Samples**	**Aggregated time in service (10^6 hrs)**		**No. of Demands**		
		Calendar time	**Operating time**			
Failure mode		**Failures (per 10^6 hrs)**			**Failures (per 10^3 demands)**	
	Lower	**Mean**	**Upper**	**Lower**	**Mean**	**Upper**
CATASTROPHIC						
a. 0 - 10% Flow Area						
b. >10% Flow Area	0.0099	0.57	2.20			
c. Rupture						
d. Plugged						
DEGRADED						
a. Restricted Flow						
INCIPIENT						
a. Wall Thinning						
b. Embrittlement						
c. Cracked or Flawed						
d. Erratic Flow						

Equipment Boundary

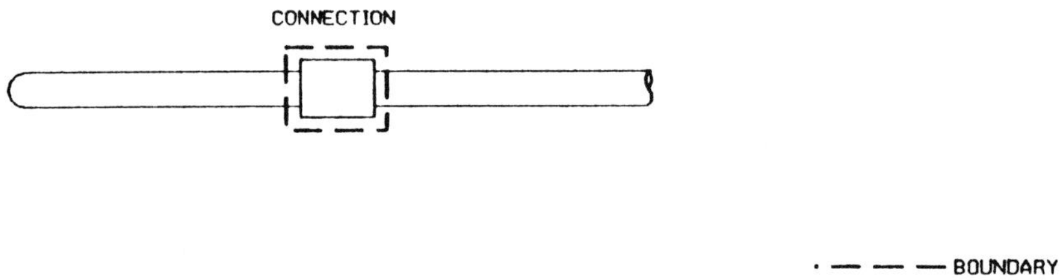

Data Reference No. (Table 5.1): 6

DATA ON SELECTED PROCESS SYSTEMS AND EQUIPMENT

Taxonomy No. 3.2.2.1	Equipment Description PIPING SYSTEMS -LINED PIPE-STRAIGHT SECTIONS
Operating Mode	Process Severity UNKNOWN

Population	Samples	Aggregated time in service (10^6 hrs)		No. of Demands
		Calendar time	Operating time	

Failure mode	Failures (per 10^6 hrs)			Failures (per 10^3 demands)		
	Lower	Mean	Upper	Lower	Mean	Upper
CATASTROPHIC	0.00743	0.442	1.71			
a. 0 - 10% Flow Area						
b. >10% Flow Area						
c. Rupture						
d. Plugged						
DEGRADED						
a. Restricted Flow						
INCIPIENT						
a. Wall Thinning						
b. Embrittlement						
c. Cracked or Flawed						
d. Erratic Flow						

Equipment Boundary

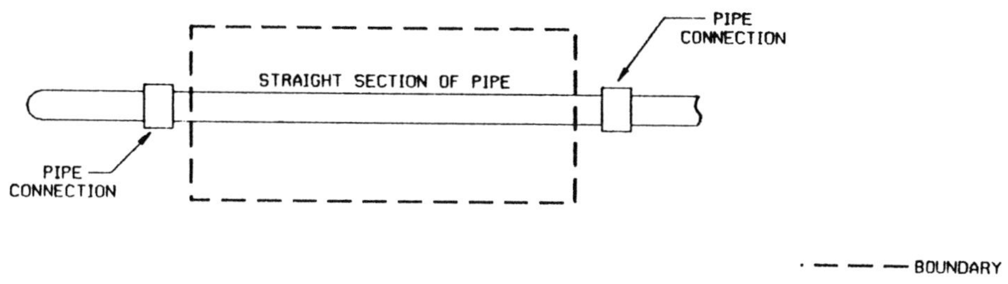

Data Reference No. (Table 5.1): 6

DATA ON SELECTED PROCESS SYSTEMS AND EQUIPMENT

Taxonomy No. 3.2.3.1	Equipment Description PIPING SYSTEMS -RIGID PLASTIC-STRAIGHT SECTIONS
Operating Mode	Process Severity UNKNOWN

Population	Samples	Aggregated time in service (10^6 hrs)		No. of Demands
		Calendar time	Operating time	

Failure mode	Failures (per 10^6 hrs)			Failures (per 10^3 demands)		
	Lower	Mean	Upper	Lower	Mean	Upper
CATASTROPHIC a. 0 - 10% Flow Area b. >10% Flow Area c. Rupture d. Plugged	0.0154	0.885	3.42			
DEGRADED a. Restricted Flow						
INCIPIENT a. Wall Thinning b. Embrittlement c. Cracked or Flawed d. Erratic Flow						

Equipment Boundary

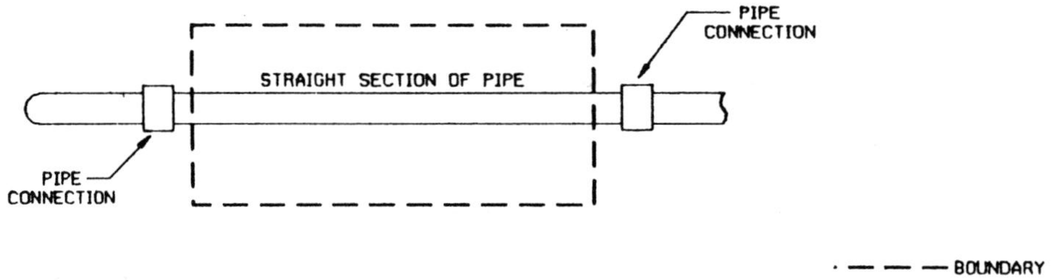

Data Reference No. (Table 5.1): 6

DATA ON SELECTED PROCESS SYSTEMS AND EQUIPMENT

Taxonomy No. 3.2.5	Equipment Description HOSES
Operating Mode	Process Severity UNKNOWN

Population	Samples	Aggregated time in service (10^6 hrs)		No. of Demands
		Calendar time	Operating time	

Failure mode	Failures (per 10^6 hrs)			Failures (per 10^3 demands)		
	Lower	Mean	Upper	Lower	Mean	Upper
CATASTROPHIC a. 0 - 10% Flow Area b. >10% Flow Area c. Rupture d. Plugged	0.0099	0.570	2.20			
DEGRADED a. Restricted Flow						
INCIPIENT a. Wall Thinning b. Embrittlement c. Cracked or Flawed d. Erratic Flow						

Equipment Boundary

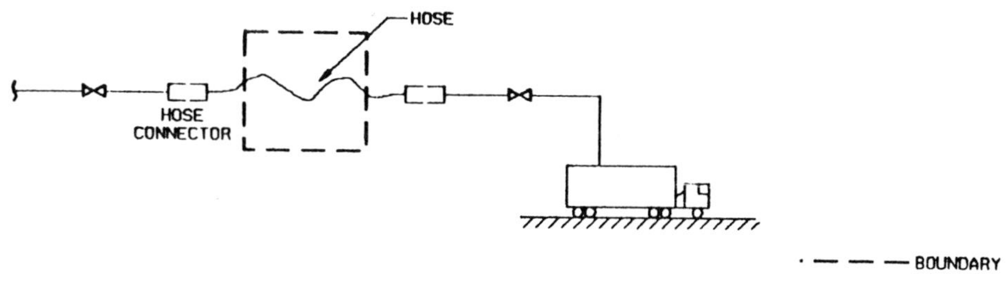

Data Reference No. (Table 5.1): 6

DATA ON SELECTED PROCESS SYSTEMS AND EQUIPMENT

Taxonomy No. 3.3.2	Equipment Description ROTATING EQUIPMENT - COMPRESSORS
Operating Mode	Process Severity UNKNOWN

Population	Samples	Aggregated time in service (10^6 hrs) Calendar time	Operating time	No. of Demands	

Failure mode	Failures (per 10^6 hrs) Lower	Mean	Upper	Failures (per 10^3 demands) Lower	Mean	Upper
CATASTROPHIC a. Fails While Running b. Rupture c. Spurious Start/Command Fault d. Fails to Start on Demand e. Fails to Stop on Demand DEGRADED a. External Leakage	3.09	1430.0	5650.0			

Equipment Boundary

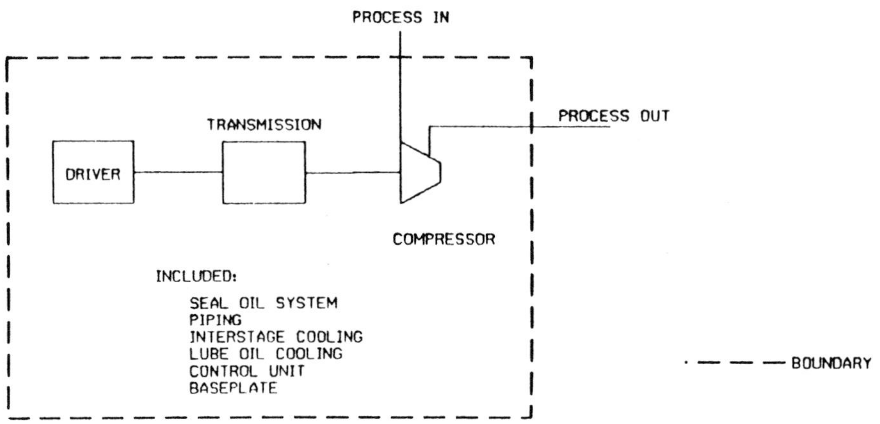

Data Reference No. (Table 5.1): 1, 5, 8.4

DATA ON SELECTED PROCESS SYSTEMS AND EQUIPMENT	
Taxonomy No. 3.3.2.1	**Equipment Description** ROTATING EQUIPMENT-COMPRESSORS-ELECTRIC MOTOR DRIVEN
Operating Mode	**Process Severity** UNKNOWN

Population	Samples	Aggregated time in service (10^6 hrs)		No. of Demands
		Calendar time	Operating time	

Failure mode	Failures (per 10^6 hrs)			Failures (per 10^3 demands)		
	Lower	Mean	Upper	Lower	Mean	Upper
CATASTROPHIC a. Fails While Running b. Rupture c. Spurious Start/Command Fault d. Fails to Start on Demand e. Fails to Stop on Demand DEGRADED a. External Leakage	27.9	2470.0	9690.0			

Equipment Boundary

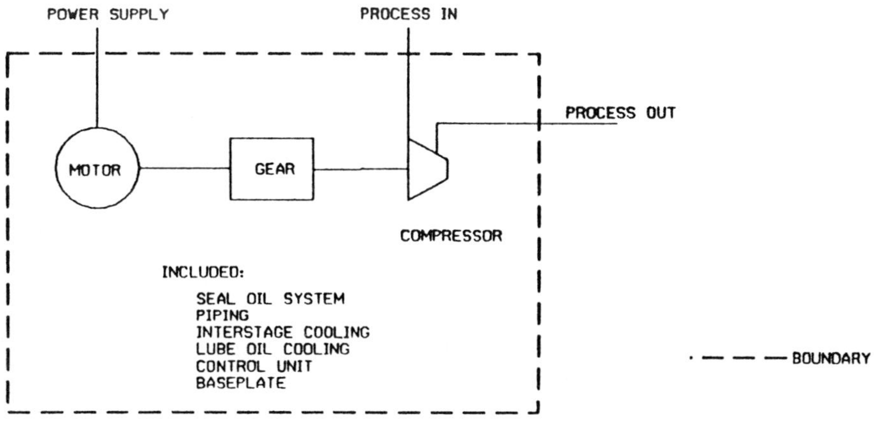

Data Reference No. (Table 5.1): 8.4

DATA ON SELECTED PROCESS SYSTEMS AND EQUIPMENT	
Taxonomy No. 3.3.2.3	**Equipment Description** ROTATING EQUIPMENT-COMPRESSORS- TURBINE-DRIVEN
Operating Mode	**Process Severity** UNKNOWN

Population	**Samples**	**Aggregated time in service (10^6 hrs)** Calendar time	Operating time	**No. of Demands**		
Failure mode	**Failures (per 10^6 hrs)** Lower	Mean	Upper	**Failures (per 10^3 demands)** Lower	Mean	Upper
CATASTROPHIC						
a. Fails While Running						
b. Rupture						
c. Spurious Start/Command Fault						
d. Fails to Start on Demand				96.5	127.0	163.0
e. Fails to Stop on Demand						
DEGRADED						
a. External Leakage						

Equipment Boundary

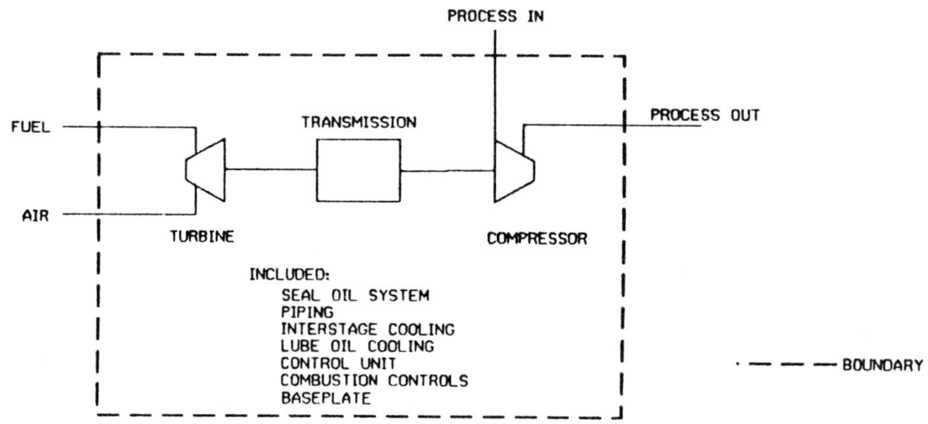

Data Reference No. (Table 5.1): 8.4

DATA ON SELECTED PROCESS SYSTEMS AND EQUIPMENT

Taxonomy No. 3.3.4	Equipment Description: ROTATING EQUIPMENT-MOTOR-DRIVEN FANS
Operating Mode	Process Severity: UNKNOWN

Population	Samples	Aggregated time in service (10^6 hrs)		No. of Demands
		Calendar time	Operating time	

Failure mode	Failures (per 10^6 hrs)			Failures (per 10^3 demands)		
	Lower	Mean	Upper	Lower	Mean	Upper
CATASTROPHIC						
a. Fails while Running	1.75	9.09	24.7			
b. Spurious Start/Command Fault						
c. Fails to Start on Demand				0.00944	0.208	0.769
d. Fails to Stop on Demand						

Equipment Boundary

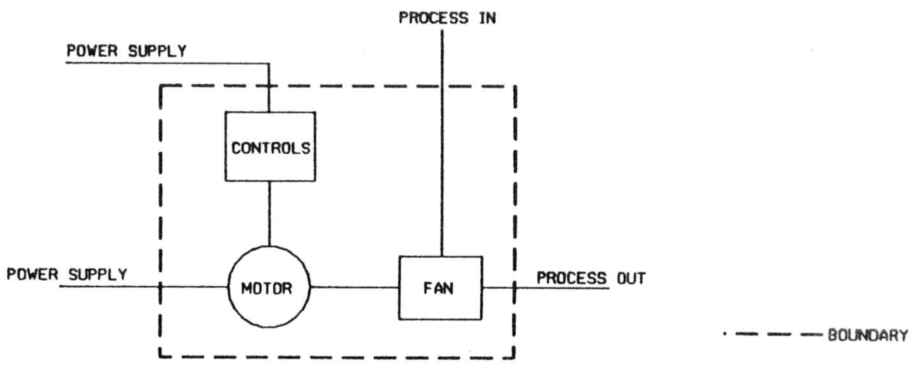

Data Reference No. (Table 5.1): 8.2, 8.4, 8.5, 8.15

DATA ON SELECTED PROCESS SYSTEMS AND EQUIPMENT

Taxonomy No.	3.3.7.2.1.1	Equipment Description	ROTATING EQUIPMENT- PUMPS MOTOR DRIVEN-PRESSURE-CENTRIFUGAL
Operating Mode	ALTERNATING	Process Severity	UNKNOWN

Population	Samples	Aggregated time in service (10^6 hrs) Calendar time	Operating time	No. of Demands

Failure mode	Failures (per 10^6 hrs) Lower	Mean	Upper	Failures (per 10^3 demands) Lower	Mean	Upper
CATASTROPHIC						
a. Fails while Running	43.3	292.0	862.0			
b. Rupture						
c. Spurious Start						
d. Fails to Start on Demand				0.360	10.80	43.0
e. Fails to Stop on Demand						
DEGRADED						
a. Fails to Run at Rated Speed	15.8	920.0	3560.0			
b. External Leak						
INCIPIENT						
a. High Vibration						
b. Over-temperature						
c. Over-current						

Equipment Boundary

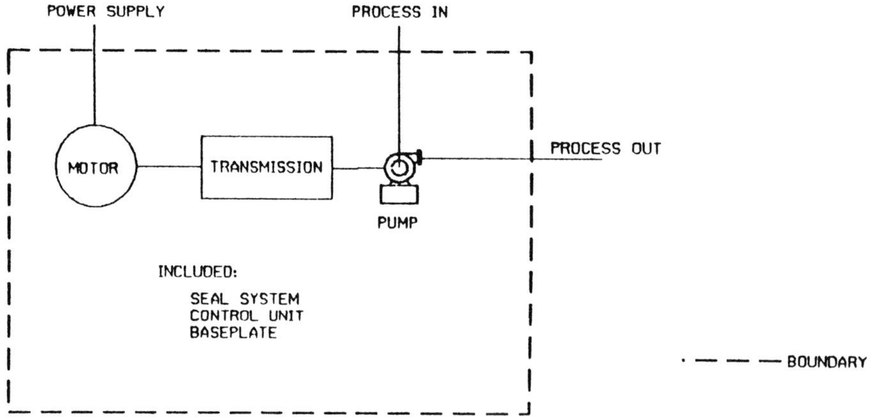

Data Reference No. (Table 5.1): 5, 8.1

DATA ON SELECTED PROCESS SYSTEMS AND EQUIPMENT

Taxonomy No. 3.3.7.2.1.1	Equipment Description ROTATING EQUIPMENT-PUMPS-MOTOR DRIVEN-PRESSURE-CENTRIFUGAL
Operating Mode RUNNING	Process Severity UNKNOWN

Population	Samples	Aggregated time in service (10^6 hrs)		No. of Demands
		Calendar time	Operating time	

Failure mode	Failures (per 10^6 hrs)			Failures (per 10^3 demands)		
	Lower	Mean	Upper	Lower	Mean	Upper
CATASTROPHIC						
a. Fails while Running	0.812	104.0	450.0			
b. Rupture						
c. Spurious Start						
d. Fails to Start on Demand						
e. Fails to Stop on Demand						
DEGRADED	0.417	24.0	92.8			
a. Fails to Run at Rated Speed						
b. External Leak						
INCIPIENT						
a. High Vibration						
b. Over-temperature						
c. Over-current						

Equipment Boundary

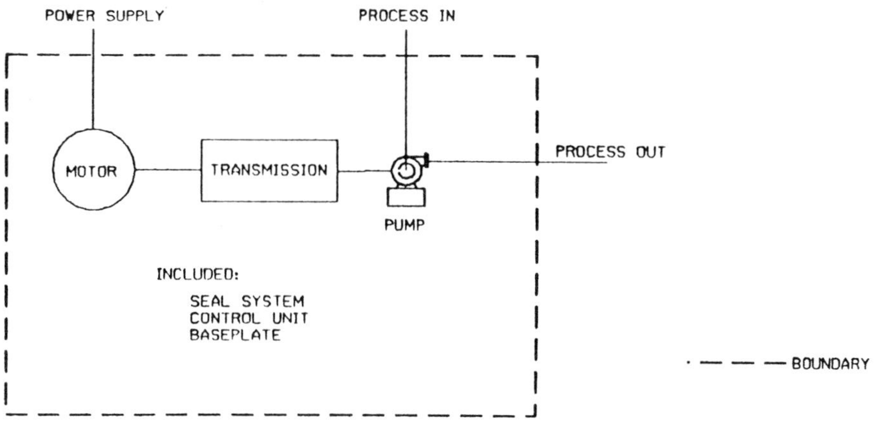

Data Reference No. (Table 5.1): 5, 8.1, 8.4

DATA ON SELECTED PROCESS SYSTEMS AND EQUIPMENT			
Taxonomy No.	3.3.7.2.1.1	Equipment Description	ROTATING EQUIPMENT- PUMPS- MOTOR DRIVEN-PRESSURE-CENTRIFUGAL
Operating Mode	STANDBY	Process Severity	UNKNOWN

Population	Samples	Aggregated time in service (10^6 hrs)		No. of Demands		
		Calendar time	Operating time			
Failure mode	Failures (per 10^6 hrs)			Failures (per 10^3 demands)		
	Lower	Mean	Upper	Lower	Mean	Upper
CATASTROPHIC						
a. Fails While Running						
b. Rupture						
c. Spurious Start						
d. Fails to Start on Demand				1.94	18.6	59.9
e. Fails to Stop on Demand						
DEGRADED						
a. Fails to Run at Rated Speed						
b. External Leak						
INCIPIENT						
a. High Vibration						
b. Over-temperature						
c. Over-current						

Equipment Boundary

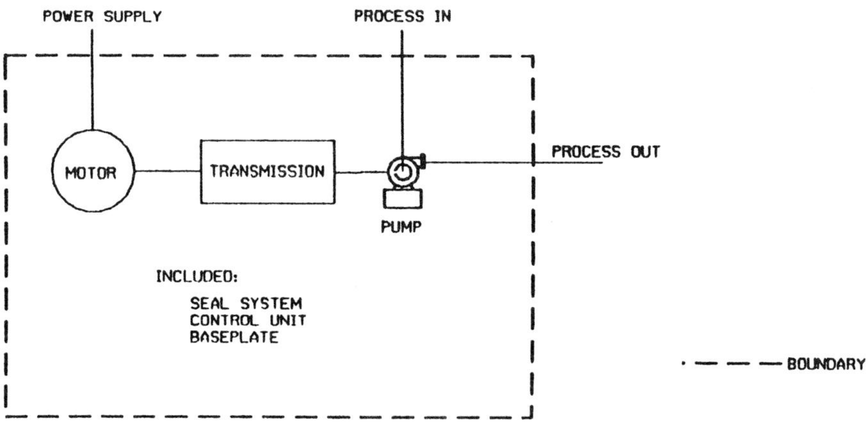

Data Reference No. (Table 5.1): 8.1

DATA ON SELECTED PROCESS SYSTEMS AND EQUIPMENT

Taxonomy No. 3.3.7.3	**Equipment Description** ROTATING EQUIPMENT-PUMPS-TURBINE-DRIVEN
Operating Mode	**Process Severity** UNKNOWN

Population	**Samples**	**Aggregated time in service (10^6 hrs)**		**No. of Demands**
		Calendar time	**Operating time**	

Failure mode	**Failures (per 10^6 hrs)**			**Failures (per 10^3 demands)**		
	Lower	**Mean**	**Upper**	**Lower**	**Mean**	**Upper**
CATASTROPHIC						
a. Fails While Running	10.9	89.1	277.0			
b. Rupture						
c. Spurious Start						
d. Fails to Start on Demand				4.18	26.2	75.8
e. Fails to Stop on Demand						
DEGRADED						
a. Fails to Run at Rated Speed						
b. External Leak						
INCIPIENT						
a. High Vibration						
b. Over-temperature						
c. Over-current						

Equipment Boundary

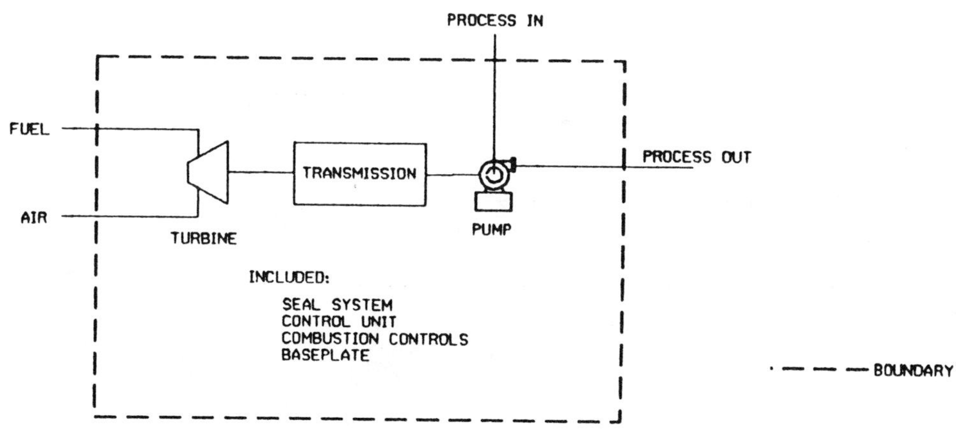

Data Reference No. (Table 5.1): 8, 8.1, 8.3, 8.7, 8.12, 8.14

DATA ON SELECTED PROCESS SYSTEMS AND EQUIPMENT

Taxonomy No. 3.4.2.2	Equipment Description SOLIDS HANDLING- CONVEYORS-SCREW
Operating Mode	Process Severity UNKNOWN

Population	Samples	Aggregated time in service (10^6 hrs)		No. of Demands
		Calendar time	Operating time	

Failure mode	Failures (per 10^6 hrs)			Failures (per 10^3 demands)		
	Lower	Mean	Upper	Lower	Mean	Upper
CATASTROPHIC	16.4	942.0	3640.0			
a. Fails While Running						
b. Rupture (External Leakage)						
c. Internal Leakage						
d. Fails to Start on Demand						
e. Fails to Stop on Demand						
DEGRADED	1.72	99.2	384.0			

Equipment Boundary

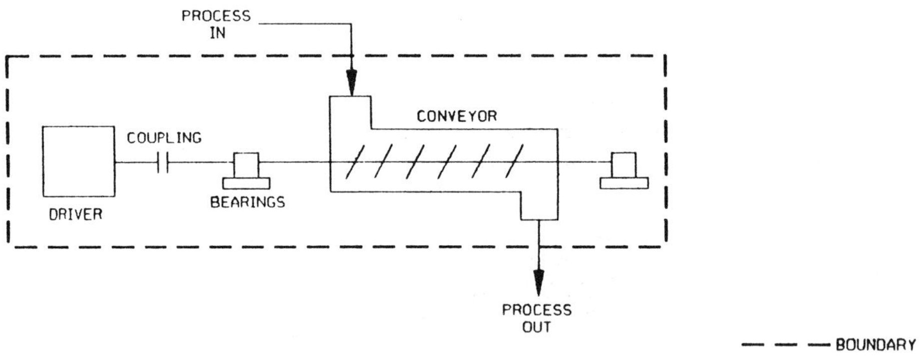

Data Reference No. (Table 5.1): 5

DATA ON SELECTED PROCESS SYSTEMS AND EQUIPMENT

Taxonomy No. 3.5.1.1	Equipment Description VALVES-OPERATED-STOP CHECK
Operating Mode	Process Severity UNKNOWN

Population	Samples	Aggregated time in service (10^6 hrs) Calendar time	Operating time	No. of Demands	

Failure mode	Failures (per 10^6 hrs) Lower	Mean	Upper	Failures (per 10^3 demands) Lower	Mean	Upper
CATASTROPHIC a. Fails to Check b. Fails to Open c. Fails to Re-open DEGRADED a. Significant Back-leakage				0.0307	1.61	6.68

Equipment Boundary

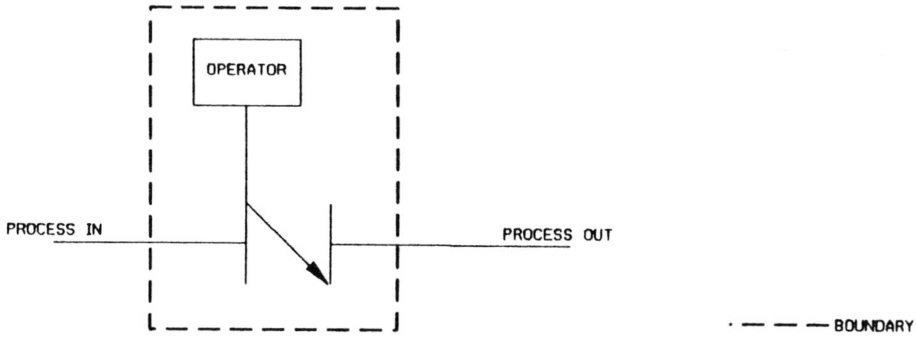

Data Reference No. (Table 5.1): 8.3, 8.12

DATA ON SELECTED PROCESS SYSTEMS AND EQUIPMENT

Taxonomy No. 3.5.1.2	Equipment Description	VALVES-NON-OPERATED-CHECK
Operating Mode	Process Severity	UNKNOWN

Population	Samples	Aggregated time in service (10^6 hrs) Calendar time	Operating time	No. of Demands

Failure mode	Failures (per 10^6 hrs) Lower	Mean	Upper	Failures (per 10^3 demands) Lower	Mean	Upper
CATASTROPHIC	0.0552	3.18	12.3			
a. Fails to Check				0.285	2.2	6.73
b. Fails to Open				0.0347	0.145	0.364
c. Fails to Re-open						
DEGRADED						
a. Significant Back-leakage						

Equipment Boundary

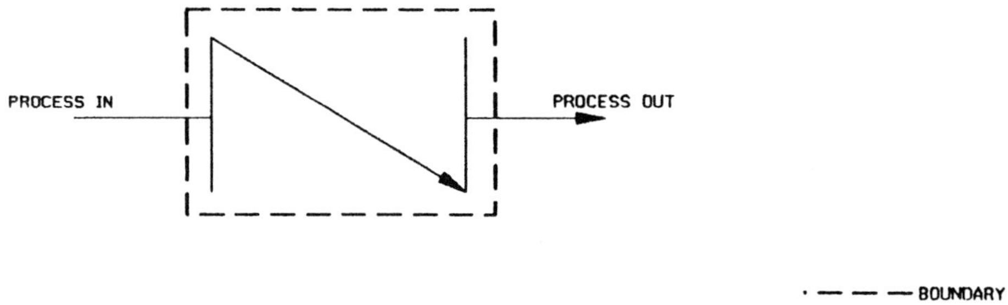

Data Reference No. (Table 5.1): 7, 8, 8.3, 8.5, 8.7, 8.11, 8.12, 8.15

DATA ON SELECTED PROCESS SYSTEMS AND EQUIPMENT

Taxonomy No. 3.5.2	Equipment Description VALVES-MANUAL
Operating Mode	Process Severity UNKNOWN

Population	Samples	Aggregated time in service (10^6 hrs)		No. of Demands
		Calendar time	Operating time	

Failure mode	Failures (per 10^6 hrs)			Failures (per 10^3 demands)		
	Lower	Mean	Upper	Lower	Mean	Upper
CATASTROPHIC	0.0141	0.152	0.501	0.0141	0.291	1.06
a. Leakage 0 - 10%						
b. Leakage >10%						
c. Rupture						
d. Normally Open/Fails Open						
e. Normally Closed/Fails Closed						
f. Normally Open/Fails Plugged						
g. Normally Closed/Fails Open						
DEGRADED						
INCIPIENT						
a. Wall Thinning						
b. Embrittlement						
c. Cracked or Flawed						
d. Internal Leakage						

Equipment Boundary

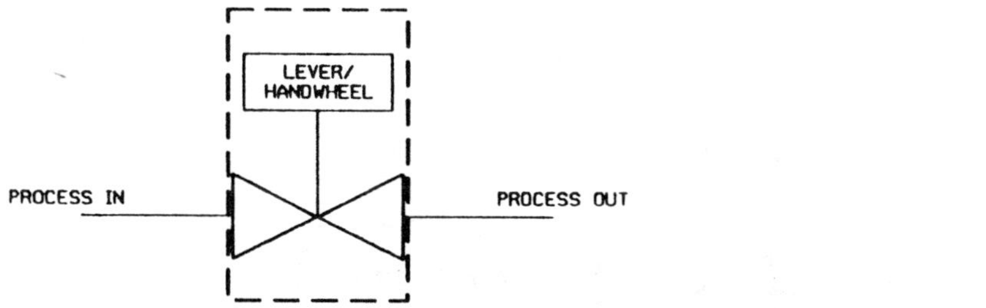

Data Reference No. (Table 5.1): 8, 8.1, 8.2, 8.3, 8.7, 8.12

DATA ON SELECTED PROCESS SYSTEMS AND EQUIPMENT

Taxonomy No. 3.5.3.2	Equipment Description VALVES-OPERATED-MOTOR
Operating Mode	Process Severity UNKNOWN

Population	Samples	Aggregated time in service (10^6 hrs)		No. of Demands	
		Calendar time	Operating time		

Failure mode	Failures (per 10^6 hrs)			Failures (per 10^3 demands)		
	Lower	Mean	Upper	Lower	Mean	Upper
CATASTROPHIC a. External Leakage b. Internal Leakage >1% c. Spurious Operation d. No Change of Position on Demand DEGRADED a. Delayed Actuation INCIPIENT a. Wall Thinning b. Embrittlement c. Cracked or Flawed d. Internal Leakage	0.239	1.36	3.80	0.504	5.58	18.6

Equipment Boundary

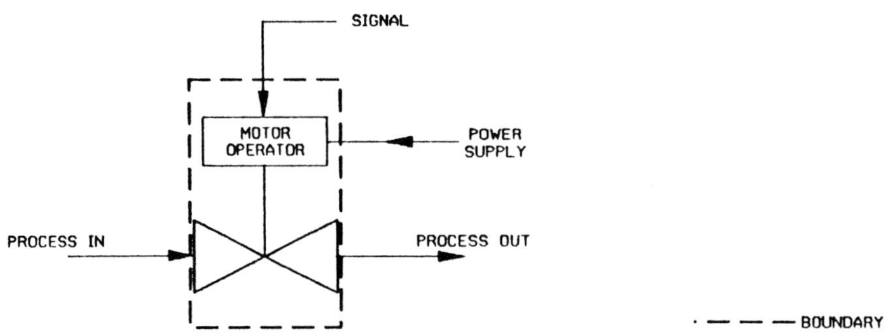

Data Reference No. (Table 5.1): 8, 8.1, 8.2, 8.3, 8.7, 8.9, 8.10, 8.11, 8.12, 8.14, 8.15

DATA ON SELECTED PROCESS SYSTEMS AND EQUIPMENT

Taxonomy No. 3.5.3.3	Equipment Description VALVES-OPERATED-PNEUMATIC
Operating Mode	Process Severity UNKNOWN

Population	Samples	Aggregated time in service (10^6 hrs)		No. of Demands		
		Calendar time	Operating time			
Failure mode	Failures (per 10^6 hrs)			Failures (per 10^3 demands)		
	Lower	Mean	Upper	Lower	Mean	Upper
CATASTROPHIC a. External Leakage b. Internal Leakage >1% c. Spurious Operation d. No Change of Position on Demand	0.274	3.59	12.3	0.306	2.2	6.62
DEGRADED a. Delayed Actuation						
INCIPIENT a. Wall Thinning b. Embrittlement c. Cracked or Flawed d. Internal Leakage						

Equipment Boundary

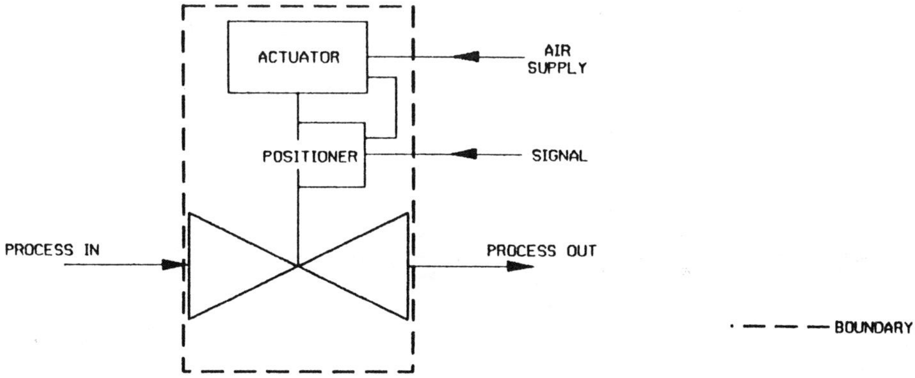

Data Reference No. (Table 5.1): 8, 8.1, 8.2, 8.3, 8.4, 8.7, 8.10, 8.12, 8.14, 8.15

DATA ON SELECTED PROCESS SYSTEMS AND EQUIPMENT

Taxonomy No.	3.5.3.4	Equipment Description	VALVES-OPERATED-SOLENOID
Operating Mode		Process Severity	UNKNOWN

Population	Samples	Aggregated time in service (10^6 hrs)		No. of Demands
		Calendar time	Operating time	

Failure mode	Failures (per 10^6 hrs)			Failures (per 10^3 demands)		
	Lower	Mean	Upper	Lower	Mean	Upper
CATASTROPHIC						
a. External Leakage	0.679	48.7	189.0			
b. Internal Leakage >10%						
c. Spurious Operation	0.108	0.409	0.985			
d. No Change of Position on Demand				0.336	2.83	10.0
DEGRADED						
a. Delayed Actuation						
INCIPIENT						
a. Wall Thinning						
b. Embrittlement						
c. Cracked or Flawed						
d. Internal Leakage						

Equipment Boundary

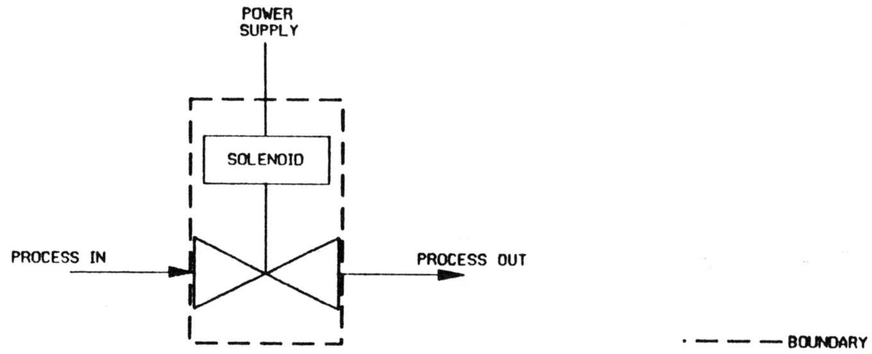

Data Reference No. (Table 5.1): 4, 7, 8, 8.2, 8.3, 8.15

DATA ON SELECTED PROCESS SYSTEMS AND EQUIPMENT	
Taxonomy No. 3.6.1.1	Equipment Description VESSELS-ATMOSPHERIC-METALLIC
Operating Mode	Process Severity UNKNOWN

Population	Samples	Aggregated time in service (10^6 hrs)		No. of Demands		
		Calendar time	Operating time			
Failure mode		**Failures (per 10^6 hrs)**			**Failures (per 10^3 demands)**		

Failure mode	Lower	Mean	Upper	Lower	Mean	Upper
CATASTROPHIC	0.127	0.985	3.02			
a. Leakage >1/4"						
b. Leakage 0 - 1/4"						
c. Rupture						
d. Plugging						
DEGRADED						
a. Restricted Flow						
INCIPIENT						
a. Wall Thinning						
b. Embrittlement						
c. Cracked or Flawed						
d. Erratic Flow						

Equipment Boundary

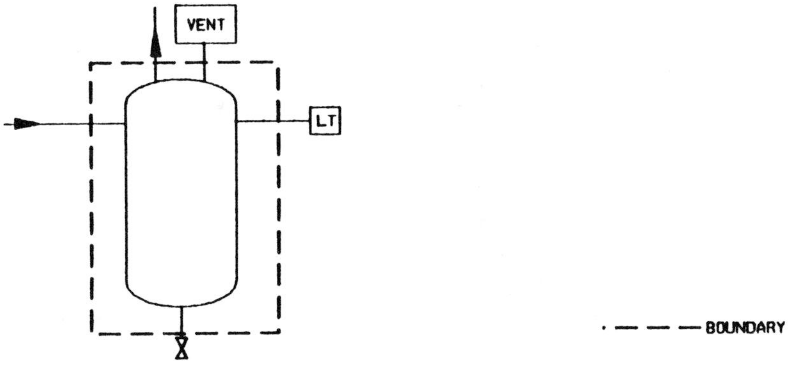

Data Reference No. (Table 5.1): 1, 6, 8.5

DATA ON SELECTED PROCESS SYSTEMS AND EQUIPMENT

Taxonomy No. 3.6.1.2	Equipment Description VESSELS-ATMOSPHERIC-NON-METALLIC
Operating Mode	Process Severity UNKNOWN

Population	Samples	Aggregated time in service (10^6 hrs)		No. of Demands	
		Calendar time	Operating time		

Failure mode	Failures (per 10^6 hrs)			Failures (per 10^3 demands)		
	Lower	Mean	Upper	Lower	Mean	Upper
CATASTROPHIC	0.021	1.21	4.68			
a. Leakage >1/4"						
b. Leakage 0 - 1/4"						
c. Rupture						
d. Plugging						
DEGRADED						
a. Restricted Flow						
INCIPIENT						
a. Wall Thinning						
b. Embrittlement						
c. Cracked or Flawed						
d. Erratic Flow						

Equipment Boundary

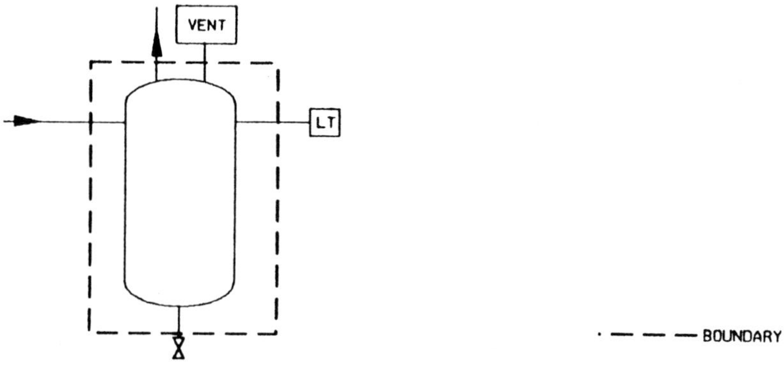

Data Reference No. (Table 5.1): 6

DATA ON SELECTED PROCESS SYSTEMS AND EQUIPMENT

Taxonomy No. 3.6.2.1	Equipment Description VESSELS-PRESSURIZED-METALLIC
Operating Mode	Process Severity UNKNOWN

Population	Samples	Aggregated time in service (10⁶ hrs) Calendar time	Operating time	No. of Demands

Failure mode	Failures (per 10^6 hrs) Lower	Mean	Upper	Failures (per 10^3 demands) Lower	Mean	Upper
CATASTROPHIC a. Leakage >1/4" b. Leakage 0 - 1/4" c. Rupture d. Plugging	0.000142	0.0109	0.0424			
DEGRADED a. Restricted Flow	0.000951	0.0636	0.247			
INCIPIENT a. Wall Thinning b. Embrittlement c. Cracked or Flawed d. Erratic Flow						

Equipment Boundary

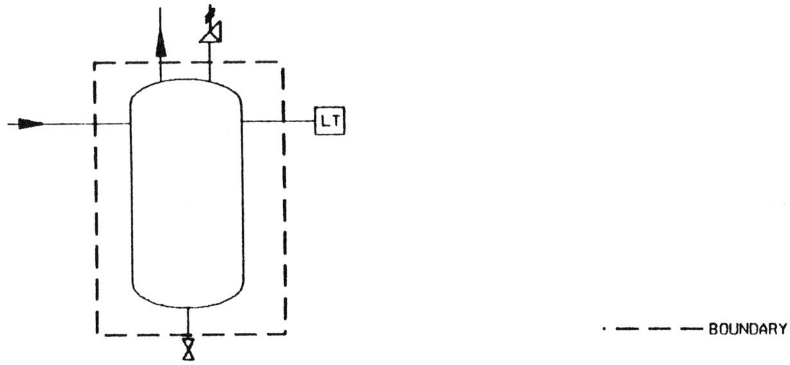

Data Reference No. (Table 5.1): 10

DATA ON SELECTED PROCESS SYSTEMS AND EQUIPMENT

Taxonomy No.	4.2.2	Equipment Description	PROTECTION SYSTEMS-FIRE-FIRE DETECTION
Operating Mode		Process Severity	UNKNOWN

Population	Samples	Aggregated time in service (10^6 hrs)		No. of Demands
		Calendar time	Operating time	

Failure mode	Failures (per 10^6 hrs)			Failures (per 10^3 demands)		
	Lower	Mean	Upper	Lower	Mean	Upper
CATASTROPHIC	0.0198	1.14	4.41			
a. Fails to Operate						
b. Spurious Operation						
c. Plugged						
d. Leaks Through						
DEGRADED						
a. Improper Operation						
b. Minor Leak Through						
INCIPIENT						
a. External Leakage						
b. Faulty Indication						

Equipment Boundary

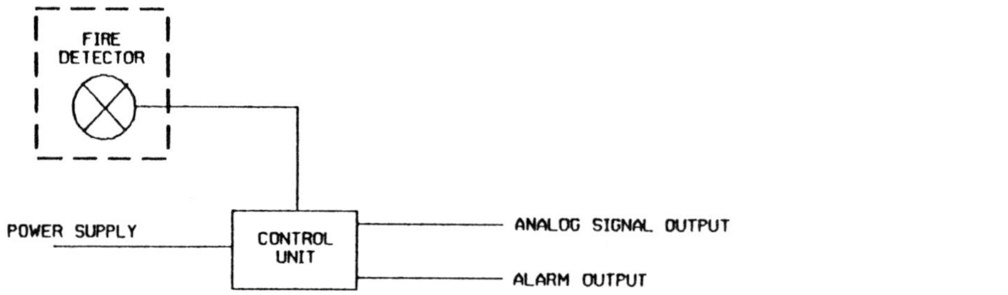

Data Reference No. (Table 5.1): 1

DATA ON SELECTED PROCESS SYSTEMS AND EQUIPMENT

Taxonomy No.	4.2.3.2	Equipment Description	PROTECTION SYSTEMS-FIRE-FIRE SUPPRESSION SYSTEMS-WATER
Operating Mode		Process Severity	UNKNOWN

Population	Samples	Aggregated time in service (10^6 hrs) Calendar time	Operating time	No. of Demands	

Failure mode	Failures (per 10^6 hrs) Lower	Mean	Upper	Failures (per 10^3 demands) Lower	Mean	Upper
CATASTROPHIC	0.168	9.66	37.4			
a. Fails to Operate						
b. Spurious Operation						
c. Plugged						
d. Leaks Through						
DEGRADED						
a. Improper Operation						
b. Minor Leak Through						
INCIPIENT						
a. External Leakage						
b. Faulty Indication						

Equipment Boundary

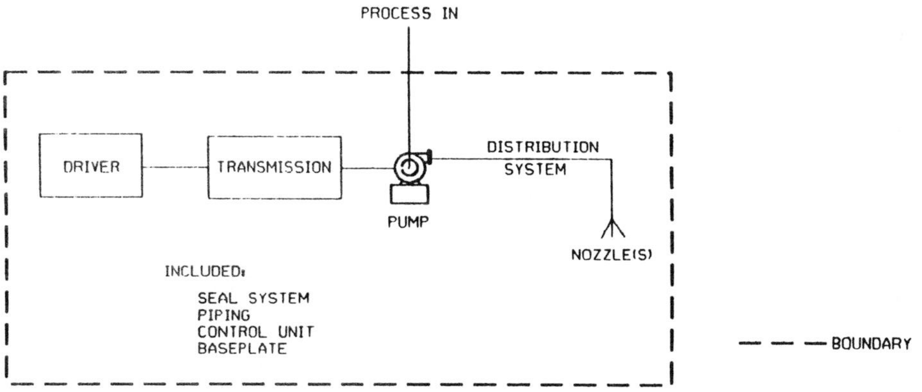

Data Reference No. (Table 5.1): 1

DATA ON SELECTED PROCESS SYSTEMS AND EQUIPMENT	
Taxonomy No. 4.2.3.3	Equipment Description PROTECTION SYSTEMS-FIRE-FIRE SUPPRESSION -DRY POWDER
Operating Mode	Process Severity UNKNOWN

Population	Samples	Aggregated time in service (10^6 hrs)		No. of Demands
		Calendar time	Operating time	

Failure mode	Failures (per 10^6 hrs)			Failures (per 10^3 demands)		
	Lower	Mean	Upper	Lower	Mean	Upper
CATASTROPHIC	0.0245	1.41	5.45			
a. Fails to Operate						
b. Spurious Operation						
c. Plugged						
d. Leaks Through						
DEGRADED						
a. Improper Operation						
b. Minor Leak Through						
INCIPIENT						
a. External Leakage						
b. Faulty Indication						

Equipment Boundary

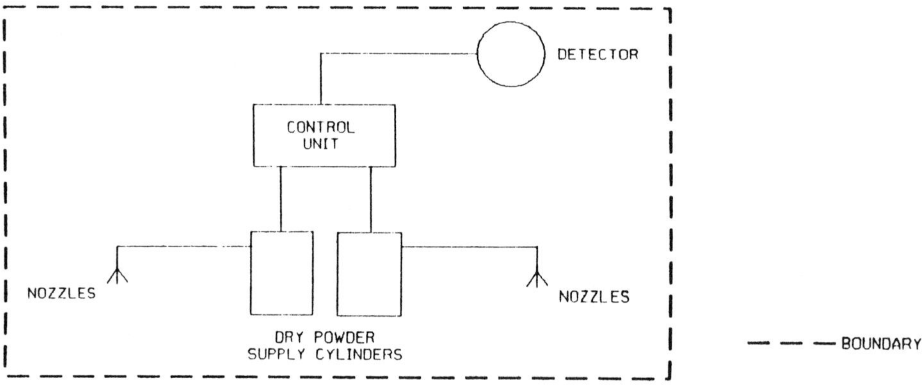

Data Reference No. (Table 5.1): 1

DATA ON SELECTED PROCESS SYSTEMS AND EQUIPMENT

Taxonomy No.	4.2.4.1	Equipment Description	PROTECTION SYSTEMS-FIRE-FIRE WATER PUMPS - DIESEL
Operating Mode	STANDBY	Process Severity	UNKNOWN

Population	Samples	Aggregated time in service (10^6 hrs)		No. of Demands
		Calendar time	Operating time	

Failure mode	Failures (per 10^6 hrs)			Failures (per 10^3 demands)		
	Lower	Mean	Upper	Lower	Mean	Upper
CATASTROPHIC						
a. Fails to Start				0.769	18.7	69.8
b. Fails While Running						
DEGRADED						
a. Low Output						
INCIPIENT						
a. Vibration						
b. Leakage						

Equipment Boundary

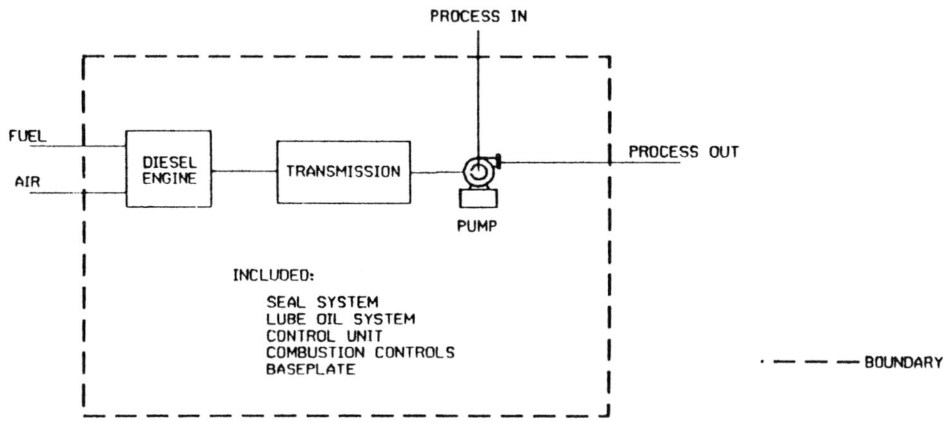

Data Reference No. (Table 5.1): 8, 8.9, 8.11

DATA ON SELECTED PROCESS SYSTEMS AND EQUIPMENT	
Taxonomy No. 4.2.4.2	Equipment Description PROTECTION SYSTEMS-FIRE-FIRE WATER PUMPS-ELECTRIC
Operating Mode	Process Severity UNKNOWN

Population	Samples	Aggregated time in service (10^6 hrs)		No. of Demands		
		Calendar time	Operating time			

Failure mode	Failures (per 10^6 hrs)			Failures (per 10^3 demands)		
	Lower	Mean	Upper	Lower	Mean	Upper
CATASTROPHIC						
a. Fails to Start				3.62	42.5	143.0
b. Fails While Running						
DEGRADED						
a. Low Output						
INCIPIENT						
a. Vibration						
b. Leakage						

Equipment Boundary

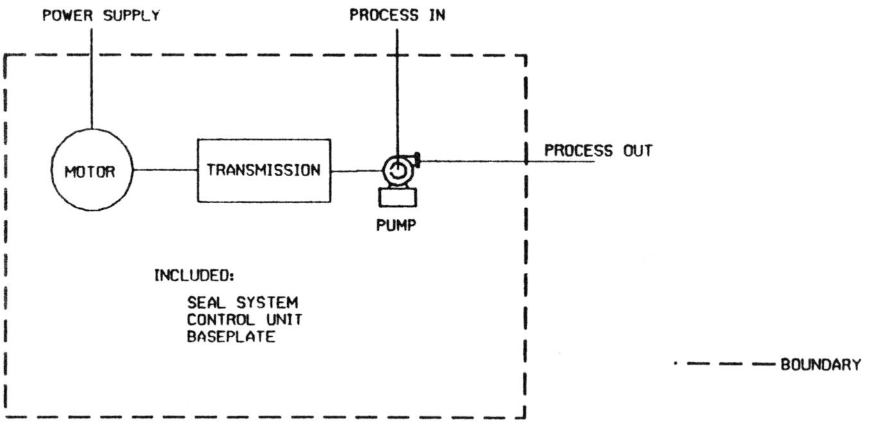

Data Reference No. (Table 5.1): 8

DATA ON SELECTED PROCESS SYSTEMS AND EQUIPMENT

Taxonomy No.	4.3.3.1	Equipment Description	PRESSURE - SAFETY RELIEF VALVES-PILOT OPERATED
Operating Mode		Process Severity	UNKNOWN

Population	Samples	Aggregated time in service (10^6 hrs)		No. of Demands
		Calendar time	Operating time	

Failure mode	Failures (per 10^6 hrs)			Failures (per 10^3 demands)		
	Lower	Mean	Upper	Lower	Mean	Upper
CATASTROPHIC						
a. Seat Leakage						
b. Fails to Open						
c. Spurious Operation						
c.1 Opens Prematurely						
c.2 Failure to Reclose once open				0.188	5.0	18.8
d. Fails to Open on Demand				0.00932	4.15	18.2
DEGRADED						
a. Interstage Leakage						
INCIPIENT						
a. Pilot Leakage						

Equipment Boundary

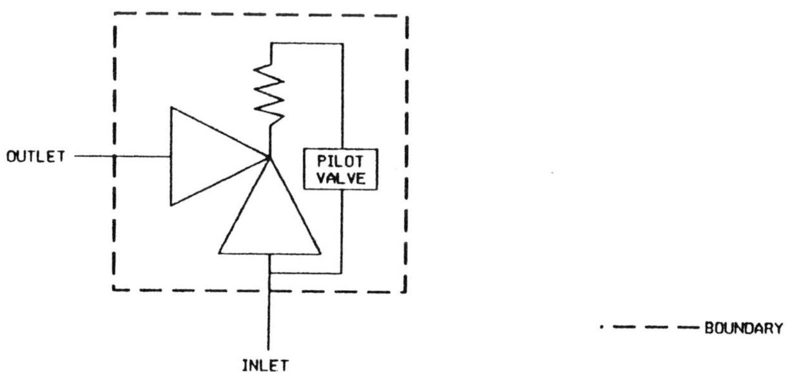

Data Reference No. (Table 5.1): 8, 8.12

DATA ON SELECTED PROCESS SYSTEMS AND EQUIPMENT

Taxonomy No.	4.3.3.2	**Equipment Description**	PRESSURE - SAFETY RELIEF VALVES - SPRING-LOADED
Operating Mode		**Process Severity**	UNKNOWN

Population	**Samples**	**Aggregated time in service (10^6 hrs)**		**No. of Demands**
		Calendar time	**Operating time**	

Failure mode	**Failures (per 10^6 hrs)**			**Failures (per 10^3 demands)**		
	Lower	**Mean**	**Upper**	**Lower**	**Mean**	**Upper**
CATASTROPHIC						
a. Seat Leakage						
b. Fails to Open						
c. Spurious Operation						
c.1 Opens Prematurely	0.275	1.68	4.80			
c.2 Failure to Reclose Once Open				0.127	5.18	22.7
d. Fails to Open on Demand				0.0079	0.212	0.798
DEGRADED						
a. Interstage Leakage						
INCIPIENT						

Equipment Boundary

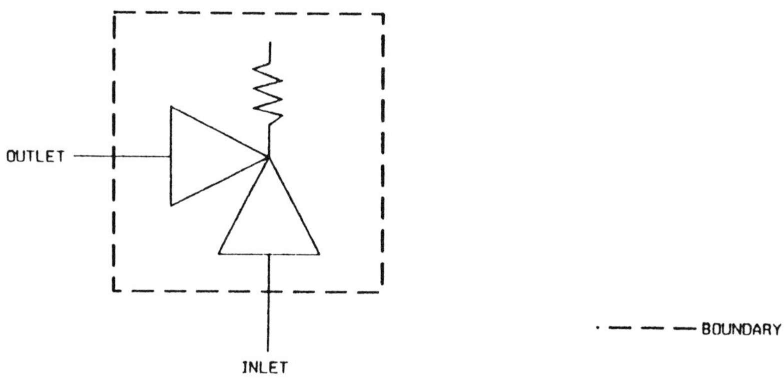

Data Reference No. (Table 5.1): 8.1, 8.3, 8.5, 8.10

6
Collection and Conversion of Plant-Specific Data

In most chemical plants, there are a wealth of data residing within various plant records. Seldom are they organized and filed in a fashion that makes them readily usable in reliability or risk analyses. The variety of forms in which maintenance, operating, and other relevant plant data are kept in different organizations makes it difficult to provide specific stepwise procedures for conversion of raw plant data. This chapter presents the thought processes and fundamentals necessary to identify the sources of data normally available and the treatment of these data to create a plant-specific failure rate data base or to add to a generic chemical industry data base. Either of these can be used to support reliability or risk analyses. This chapter is also intended to help the reader develop record keeping systems that will provide useful, pertinent failure rate data for risk analyses and yield benefits beyond operating and maintenance requirements.

It should be noted that the data collection and conversion effort is not trivial, it is company and plant-specific and requires substantial effort and coordination between intracompany groups. No statistical treatment can make up for inaccurate or incomplete raw data. The keys to valid, high-quality data are thoroughness and quality of personnel training; comprehensive procedures for data collection, reduction, handling and protection (from raw records to final failure rates); and the ability to audit and trace the origins of finished data. Finally, the system must be structured and the data must be coded so that they can be located within a well-designed failure rate taxonomy. When done properly, valuable and uniquely applicable failure rate data and equipment reliability information can be obtained.

6.1 Data Sources

Rates of equipment failure are calculated by dividing the number of failures for an equipment population by its total exposure hours or total number of demands. The following key types of information, therefore, are needed to develop plant-specific failure rate data:

- population of basic types of equipment;
- number of equipment failures, classified by failure mode;
- equipment exposure time (both calendar and operating time) and demands as applicable.

6.1.1. Equipment Population

The first step in failure rate determination for a specific type of process equipment is to obtain a list of the installed equipment at the plant and sufficient description data to assign it a number within the CCPS Taxonomy. The descriptive information desired by the analyst is:

- physical equipment description;
- equipment boundary definition;
- service description; and
- installation and process environment.

Although maintenance systems contain some of this information, engineering, purchasing, and operating department records may be required to find the remainder. Also, equipment maintenance records may be in several file locations since they are usually organized by components and component modules that may differ from equipment boundaries established for risk analysis.

Ideally, maintenance records should be organized by a classification method compatible with the CCPS Taxonomy in Appendix A and the equipment boundaries in Section 5.5, Generic Failure Rate Data Base. It is important to remember that the taxonomy presented was developed to group equipment into classes that are differentiated by their reliability rather than their design characteristics. Records maintained in this fashion allow the analyst to more easily determine the total pieces of equipment and number failures.

6.1.2. Equipment Failures

Within the maintenance system, there is generally a means to determine when maintenance work is required and when that work is completed. Different facilities call such records by different names, among them: work orders, work requests, trouble tickets, maintenance records, and work authorizations. Despite the variation in names, this valuable information contains:

- Date of issue (date failure was noted and documented)
- System and/or equipment identification number affected
- Description of failure/condition of equipment observed
- Description of corrective action taken
- Date and sign-off of completion.

The records display a pattern of maintenance and repair that is rarely visible elsewhere and can show less severe equipment damage trends that can lead to total failure. As such, it is possible to determine the total number of failures and failure severity.

Other reports used within facilities record failures of particular interest because of failure mode and system or equipment affected. Some facilities may issue special reports when the plant experiences a shutdown (outage report) or when the occurrence is sufficiently unique or troublesome to warrant further investigation (unusual event report). In general, these reports can be characterized by their relatively restrictive focus (when compared to the maintenance records) and their smaller number.

6.1.3. Equipment Exposure Time

The number and severity of failures experienced by the equipment under study must be related to the operations of the facility. It would be inappropriate to assign the same operating histories to a continuously operating system and a system that operates intermittently. The number of hours in different operating modes (for example, 100% production versus shutdown) affect failure rate calculation and service description for taxonomy definition.

Most facilities keep records on the operating status of the plant, usually in the form of monthly status reports or a chart that displays production level versus days, weeks, or months over the plant life. Changes in plant status are generally noted by date on either of these two data sources, but may also be logged separately. This information is important so that an accurate count of the number of hours spent in each plant state (operating versus nonoperating) and number of demands due to plant state changes can be used for reliability and risk analysis.

Testing and periodic maintenance can place additional demands on systems, equipment, and components that render them unavailable for service. These details must also be factored into the calculation of operating time for a piece of equipment.

6.2 Data Collection

Once it is determined that data exist, the next step is to begin the collection process. If sufficient thought and training is provided in the development and operation of the maintenance and operating reporting systems, much of the collection process can be automated. Automation assumes that a well-thought-out taxonomy is in place. If this is not the case, then an analyst must collect and review the records manually. In either case, the analyst must collect data from the plant sources previously discussed in order to determine the numerator (number of failures within a unique plant equipment population), and denominator (the operating time or number of demands for the equipment) of the equation to calculate failure rates.

6.2.1 Data Collection Procedures

Data collection procedures must be established to capture the required information. Various methods have been proposed for collecting such data, including a draft international standard that provides criteria for collecting data in nuclear power plants.[1] These criteria are also useful in developing methodologies for collection of data within the CPI. Smith and Babb[2] provide additional information beyond that presented in this book.

Basic data collection procedures need to be comprehensive and formalized. They should address completion of the collection forms, the filing and distribution of these forms, and retention requirements of data source materials and other documents.

Documentation of data origin is essential. Each completed data collection form needs to contain a file reference number or code to connect it to the documentation sources. This provides an essential trail to audit data quality, to confirm risk or reliability estimates or to investigate data values that appear questionable. Procedures to control data during handling, processing, recording, and reviewing are also necessary to prevent loss of data and to assure that opportunities are not lost to check the content of a form, by complete or random sampling. These checks on data quality can be accomplished by computerized audits or manually by the preparer and/or "data auditor." Bendell[3] suggests that data collection errors may affect 10–20% of the data collected and recommends that the problem be reduced by careful data validation.

The individual responsible for completing the data input forms needs formalized training in data collection procedures, with written instructions on form completion, on data handling, and on documentation procedures. Other material needed to encode raw data properly must be available. In addition, these individuals need access to a consultant within the organization to help resolve questions that may arise.

If reviewers of completed forms are not the preparer, they need to be trained in procedures to audit the quality of the collected information and documentation files. This training may include means to check the completeness and credibility of the collected data by cross checking the data against other reference files, such as maintenance files or operating logs.

Following the review, audit, and acceptance process, the preparer needs to be notified of data acceptance or rejection. With this feedback, the preparer can make necessary corrections to provide higher quality data in the future.

6.2.2 Data Collection Forms

Given the large variety of maintenance and operating systems and procedures that exist within the CPI, it is impossible to provide a set of data collection forms that will completely satisfy the requirements of all users. However, sample forms are included that contain many of the elements that data collection efforts must address. Using these as a reference, the reader should be able to capture necessary raw data.

Various forms have been developed to collect plant data. Figures 6.1 and 6.2 are generic forms published in EuReDatA Project No. 3.[4] The Specimen Inventory form, Figure 6.1, is designed to collect data needed to establish the equipment description and total equipment population. Many maintenance systems offer some of these data, but usually not in a useful format or to the extent desired. The Specimen Event or Failure Report form, Figure 6.2, is used to capture failure event data that, when summed, will allow determination of the failure rate numerator—the number of failures within a unique plant population.

A different set of forms, in extensive use for failure rate calculation, are used to illustrate the remaining sections of this chapter. Beginning with Figure 6.3, the forms present a worked pump example for the conversion of actual plant raw data to plant-specific failure rate data.

6.3 Data Review and Qualification

It is important, especially when consistency has not been designed or built into the maintenance reporting system, to review the data reported to minimize misinterpretation. Clearly defined equipment boundaries for plant hardware are essential for the generation of relevant data. For example, one classification method may define pumps as only the mechanical portions of the pump, whereas another may include the driver (e.g., the motor) and associated controls. Interviews with operating and maintenance personnel as well as review of the maintenance procedures and documents can provide insight into the

Specimen Inventory Form

Inventory No. ....
Related Inventory No. ....

Equipment Description ....
....
....

No. in Population ....
Plant Identification No. ....
Information Source ....

Manufacturers name ....
Manufacturers Model No. ....
Manufacturers Serial No. ....

Generic Family Type ....
Operating Duty ....
Environmental Conditions ....
....

Medium in use ....
Constructional Material ....
Design Specifications ....

Maintenance & Test Types
BREAKDOWN
SCHEDULED — Maintenance ....
UNSCHEDULED
OPPORTUNE — Test ....

Maintenance Period .... Test Period ....

Design Parameters

Value	Sig	Units	Description
....	....	....	....
....	....	....	....
....	....	....	....
....	....	....	....

Working Parameters

Value	Sig	Units	Description
....	....	....	....
....	....	....	....
....	....	....	....
....	....	....	....

Comments ....
....
....

Completed By (print) .... Date --/--/----

Approved By (print) .... Date --/--/----

Figure 6.1 *Specimen Inventory Form—EuReDatA Project No.3. From EuReDatA Project No. 3, Guide to Reliability Data Collection and Management.*

Specimen Event or Failure Report Form

Event No. ....
Equipment Serial No. ....
Equipment Description ....
....
Plant Identification No. ....
Location ....
Date installed at location --/--/----
Condition on installation NEW/REFURBISHED/ ....

Time and date of failure .... --/--/----
Failure Description ....

Method of observation
ALARM/TRIP
SCHEDULED TEST/MAINTENANCE
OPPORTUNE TEST/MAINTENANCE
OTHER (specify) ....

Equipment status on failure
ON STANDBY (HOT/COLD)
FULL LOAD 80%
PARTIAL LOAD 80%
EXCESS LOAD 100%
SHUTDOWN
OTHER (specify) ....

Plant status on failure ....

Method of repair
REPAIR IN SITU
REMOVE TO WORKSHOP — REPAIR AND REPLACE IN ORIGINAL LOCATION
REMOVE TO WORKSHOP — REPAIR AND STORE
SCRAP..

Replacement Equipment Serial Number ....
Replacement Equipment Condition NEW/REFURBISHED/ ....

Time and date repair/replacement started .... --/--/----
Time and date repair/replacement completed .... --/--/----
Time and date equipment location functional .... --/--/----

Details of repair (reports from each trade giving actions taken and times)

Trade	Report	Times (hrs
....	....	....
....	....	....
....	....	....
....	....	....

Related Equipment Event Report Nos ....
Failure Diagnosis Report ....
....

Failure Mode .... Failure Cause ....

Comments ....
....

Completed by (print) .... Date --/--/----

Approved by (print) .... Date --/--/----

Figure 6.2 *Specimen Event or Failure Report Form—EuReDatA Project No.3. From EuReDatA Project No. 3, Guide to Reliability Data Collection and Management.*

uniformity of equipment boundary definition and the quality and consistency of the data recorded according to these boundaries.

6.4 Data Conversion

After collecting the necessary information from the plant's files and employees, it is crucial to have a structured approach to reduce and combine the raw data into a relevant form for analysis and failure rate computation. The following steps describe these procedures:

- Establishment of the study time frame
- Data encoding
- Compilation of failure severities
- Determination of operating hours/equipment demands
- Computation of failure rates.

6.4.1. Establishment of Study Time Frame

The recommended initial step is to determine the duration of the study period for the analysis by first defining a "start date." The "end date" is determined by the maintenance records. For computerized systems, the last update of the overall system provides the end date. For manual systems, it may be necessary to define an end date to accommodate the lag time in updating records. The start and end dates define the maximum calendar time that a piece of equipment is available for service. The operating time for this equipment will be equal to or less than the calendar period, depending on plant operations and the equipment's operating mode.

6.4.2. Data Encoding

Data encoding consists of converting historical failure records into formats that can then be used to calculate time-related and demand-related equipment failure rates. It is important that the encoding step be done so that plant information used to calculate failure rates can be traced through each step back to the raw data. The minimum set of information that must be recorded includes:

- Date of occurrence
- Data source identification
- System identification
- Equipment type
- Equipment identification
- Failure mode and severity
- Time-related or demand-related failure determination

An example of a form used to collect and structure the encoded information is shown as Figure 6.3.

The date of occurrence is an important element since it verifies whether the equipment failure occurred during the analysis time frame. The equipment type is necessary for

Pumps

PLANT: N-Rx SYSTEM: #37 Primary

DATE 3-14-84
RECORD # 127735
COMP ID# PCP-6
COMP. TYPE
FAIL SEVERITY D
FAIL MODE
FAIL CAUSE CODE
Demand or Time T

DATE 3-25-83
RECORD # 120738
COMP ID# PCP-6
COMP. TYPE
FAIL SEVERITY I
FAIL MODE
FAIL CAUSE CODE
Demand or Time T

DATE 1-23-85
RECORD # 133369
COMP ID# PCP-4
COMP. TYPE
FAIL SEVERITY I
FAIL MODE
FAIL CAUSE CODE
Demand or Time T

DATE 1-6-84
RECORD # 126641
COMP ID# DCP-4
COMP. TYPE
FAIL SEVERITY D
FAIL MODE
FAIL CAUSE CODE
Demand or Time T

DATE 9-17-84
RECORD # 131168
COMP ID# PCP-5
COMP. TYPE
FAIL SEVERITY I
FAIL MODE
FAIL CAUSE CODE
Demand or Time T

DATE 6-2-76
RECORD # 082192
COMP ID# PCP-5
COMP. TYPE
FAIL SEVERITY I
FAIL MODE
FAIL CAUSE CODE
Demand or Time T

DATE 11-2-83
RECORD # 125660
COMP ID# PCP-1
COMP. TYPE
FAIL SEVERITY C
FAIL MODE
FAIL CAUSE CODE
Demand or Time Demand

DATE 9-17-84
RECORD # 131171
COMP ID# PCP-2
COMP. TYPE
FAIL SEVERITY I
FAIL MODE
FAIL CAUSE CODE
Demand or Time T

DATE
RECORD #
COMP ID#
COMP. TYPE
FAIL SEVERITY
FAIL MODE
FAIL CAUSE CODE
Demand or Time

DATE 11-22-83
RECORD # 126018
COMP ID# PCP-6
COMP. TYPE
FAIL SEVERITY D
FAIL MODE
FAIL CAUSE CODE
Demand or Time T

DATE 9-17-84
RECORD # 131172
COMP ID# PCP-6
COMP. TYPE
FAIL SEVERITY I
FAIL MODE
FAIL CAUSE CODE
Demand or Time T

DATE
RECORD #
COMP ID#
COMP. TYPE
FAIL SEVERITY
FAIL MODE
FAIL CAUSE CODE
Demand or Time

DATE 7-16-84
RECORD # 120224
COMP ID# PCP-6
COMP. TYPE
FAIL SEVERITY D
FAIL MODE
FAIL CAUSE CODE
Demand or Time T

DATE 4-15-85
RECORD # 135241
COMP ID# PCP-6
COMP. TYPE
FAIL SEVERITY I
FAIL MODE
FAIL CAUSE CODE
Demand or Time T

DATE
RECORD #
COMP ID#
COMP. TYPE
FAIL SEVERITY
FAIL MODE
FAIL CAUSE CODE
Demand or Time

Figure 6.3 *Example Data Encoding Form. From Science Applications International Corporation.*

proper placement in the CCPS Taxonomy, whereas the equipment, system, and data identification provide traceability back to the raw data. By using the equipment identification and the date of occurrence together, it is possible to identify duplicate records. Assignment of failure cause code is recommended at this stage for reliability engineering analysis.

Sometimes it is necessary to review the narrative in raw data records to determine whether a failure has occurred, to establish failure modes and severities, and to see if a record is a duplicate or new failure. Often, the narrative section is the only way the data analyst can determine if the document, especially a work order, is for a legitimate failure, routine maintenance, or a specified test.

In the event that maintenance-related unavailabilities need to be determined from the work requests, the following should also be encoded at this time:

- Duration equipment rendered inoperable
- Duration subsystem rendered inoperable
- Duration system rendered inoperable

A complete understanding of the failure process is necessary to determine if a recordable failure has occurred, the mode of the failure, and the equipment to which the failure is assigned. This is not always obvious from some plant records. The following examples illustrate this problem:

> An operator, while performing a normal test of a pump, attempts to start the pump using the switch on the control room panel. The pump fails to start after repeated tries. The operator initiates a work order specifying the pump has failed. During troubleshooting, the repair technicians discover that the pump controller's (referred to as the circuit breaker) starting resistor was improperly replaced during recent preventive maintenance and prevented the pump from starting. Although human error is the initiating failure cause, this is catastrophic equipment failure, because the pump did not start on demand. Consequently, the equipment involved is charged with this demand-related catastrophic failure.
>
> If the circuit breaker is excluded from the pump equipment boundary and is defined as part of the controller that includes the circuit breaker, starting circuit, and protective trip circuits, then the circuit breaker has failed.
>
> In the CCPS taxonomy, where the equipment boundary for a motor-operated pump includes the motor, shaft, seals, casing, impeller and the circuit breaker, then the demand-related failure is attributed to the pump.

Due to such subtleties, the need to develop well-defined basic events, failure modes, and equipment boundaries prior to data encoding cannot be overemphasized. Familiarity with failure definitions and failure severities will be extremely helpful to the analyst. Figures 2.1 and 2.2, reprinted from IEEE Std. 500-1984[5], list a large number of failure modes by failure severity and may help encode failures. IPRDS[1] also contains helpful information on failure encoding. Information on some equipment boundaries may be found in the Data Tables in Section 5.5.

The primary value of proper and consistent data encoding is that it preserves the quality of input data. In addition, the use of accepted standard encoding schemes provides data that can be compared and combined with generic data.

6.4.3. Compilation of Failure Severities

When the encoding process has been completed, the number of failures can be compiled for each failure severity through the use of a Failure Compilation Worksheet such as the one shown in Figure 6.4. The total failures in each failure severity category are noted with subtotals in brackets, where the left-hand subtotal denotes demand-related failures and the right-hand subtotal provides the time-related failures. These subtotals within the brackets are separated by placing a comma (,) between them. The total number of catastrophic failures for each type of equipment and type of failure (demand or time) provides the numerator for the catastrophic failure rates. It should be noted that failure data for lower severity failure modes (degraded and incipient) should use the totals in the appropriate severity category as their numerators.

6.4.4. Determination of Operating Hours/Equipment Demands

When the number of failures has been determined for a specific type of equipment in a particular system, the next step is to estimate the total operating hours for each equipment type in that system and/or the total number of demands on that equipment. A method for estimating these parameters is required unless operating hours are logged and demands are accounted for and clearly documented. The actual logging of operating hours and demands has traditionally been rare. For this reason, a complete understanding of system information, including operating characteristics, test information, and plant operating history is required to make informed estimates of exposures.

The time-related exposure corresponds to the historical operating time of the equipment population. It is generally necessary to use an indirect method for estimating equipment operating times. The actual plant records reviewed provide the basis for estimating exposure time. The operating history of the plant is reduced to total operating hours and shutdown hours, the total number of startups and shutdowns, and the operating mode and experience of each piece of equipment. This information is then summarized and used for

Total [demand, time]

DATA WORKSHEET - PUMP

SYSTEM : PRIMARY COOLING #37

PUMP ID *	PCP #1	PCP #2	PCP #3	PCP #4	PCP #5	PCP #6
CATASTROPHIC	5 [1, 4]	1 [0, 1]	3 [1, 2]	6 [0, 6]	2 [0, 2]	2 [0, 2]
DEGRADED	4 [0, 4]	1 [0, 1]	3 [0, 3]	5 [0, 5]	3 [0, 3]	14 [0, 14]
INCIPIENT	13 [0, 13]	19 [0, 19]	13 [0, 13]	26 [0, 26]	32 [0, 32]	25 [0, 25]
TOTAL	22 [1, 21]	21 [0, 21]	19 [1, 18]	37 [0, 37]	37 [0, 37]	41 [0, 41]

* Includes Drive Turbine

Figure 6.4 *Example Failure Compilation Worksheet Note: The format used to enter failure data is X[Y,Z],* where *X* is the total number of failures, *Y* is the number of demand-related failures, and *Z* is the number of time-related failures. From Science Applications International Corporation.

PUMP TIME RELATED FAILURE DATA SHEET

SYSTEM : PRIMARY COOLING

PUMP TYPE : PRIMARY COOLANT (PCP) DRIVER TYPE : TURBINE

TOTAL EXPOSURE TIME : (T) 64 174 hours for each PCP

CRITERIA USED FOR DETERMINING EXPOSURE TIME (T) :

Exposure time: 5 out of 6 PCPs run during normal operation; 2 to 5 PCPs may be running during Shutdown (S/D); Assume 3 PCPs for calculational purposes [these assumptions obtained from systems analysts via plant drawings and discussions w/operators].

Exposure time = 5/6 × (47874 hrs) + 3/6 × (48558) = 39895 + 24279

5 out of 6 pumps (normal operation) 3 out of 6 (S/D hrs)

Calculated from plant history records

= 64174 hours

PUMP ID	PCP#1	PCP#2	PCP#3	PCP#4	PCP#5	PCP#6
TOTAL # OF TIME DEPENDENT FAILURES	4	1	2	6	2	2
FAILURE RATE	6.23×10^{-5}	1.56×10^{-5}	3.12×10^{-5}	9.35×10^{-5}	3.12×10^{-5}	3.12×10^{-5}

Figure 6.5 *Example Pump Time-related Failure Rate Data Worksheet. From Science Applications International Corporation.*

the failure rate calculations. Figure 6.5 illustrates a data sheet that can be used to log the information necessary to compute the estimated time-related failure rate.

The demand-related exposure is the historical number of demands for change of state experienced by the equipment population. In theory, the evaluation of this demand exposure is rather simple. In practice, the number of demands on a piece of equipment can originate from four different sources: testing, automatic and manual initiation, failure related maintenance, and interfacing maintenance. The contribution each of these categories makes to demand related exposures is summarized below:

Test demands: Periodic equipment testing is an important source of demands, especially for safety system components that are often in a standby state. A review of the test procedures can be performed to obtain this information if it is not recorded in the maintenance records.

Automatic and manual initiation demands: In addition to test demands, equipment may be activated or deactivated by intended or spurious signals. These signals include losses of offsite power and normal shutdowns or startups. Different components react to different signals or sets of signals, depending on their functions for each system. The information on initiating signals can be obtained from maintenance work requests or operating logs.

Failure related maintenance demands: Whenever a repair is made on a piece of equipment, a complete functional check-out is usually performed on it and the other equipment in the functional loop. Consequently, the number of shutdown and startup demands to repair degraded and incipient failures must be added to the number of catastrophic demand failures to determine total maintenance-related demands. These numbers are gathered from the encoded failure data that are in turn extracted from the raw plant records.

Interfacing maintenance demands: When maintenance is completed on equipment, for example, a pump, a complete functional check of it and its associated equipment (e.g., isolation valves), is usually performed before the equipment is returned to service. It is important to note that in this case the interface demand is placed on the associated equipment, not the equipment that failed. Interfacing maintenance demands are counted separately from failure-related demands.

The following provides an example of the steps to estimate interfacing maintenance demands for the case above:

- List equipment requiring isolation (pump).
- Record the number and types of associated equipment (valves) that interface with the equipment requiring isolation.
- Calculate the total quantity of each isolation valve type.
- Group similar types of isolation valves together according to the taxonomy.
- Calculate interface demands on the isolation valves.

The information from these four sources of demands are then summarized and used for the failure rate calculations. Demand-related data sheets, appropriate to the hardware (Figures 6.6, 6.7, and 6.8) can be used to log information to compute demand-related failure rates.

6.4.5 Computation of Failure Rates

When the numerator data have been established (through the encoding of equipment failure records) and the denominator data have been developed (through exposure and demand calculations), failure rates can be computed. An example of this is shown in Figure 6.5, which contains the data and results of time-related failure rate calculations. Figures 6.6 and 6.8 provide examples of the data and calculations for demand-related failure rates. The rates of failure can be summarized by major equipment types on a Rate of Failure Summary form, such as the example for pumps in Figure 6.9.

6.4.6 Incomplete Information

The data presented in Figure 6.10 illustrate the need for complete information. As it stands, presented as failures per 10^6 hours, it is useless. These data are based on the total number of failures in 14.5 years of service for a system with an unknown number of pipe fittings. To give the data value, the number of fittings in the system must be known in order to derive usable failure rates per 10^6 hours *per fitting*.

VALVE DEMAND DATA SHEET

SYSTEM : PRIMARY COOLING #37

VALVE TYPE

Population	MT	MG	RV		
DEMANDS	20	5	2		
TEST	100	25	40		
AUTO. & MAN.	0	0	20		
FAILURE RELATED	43	20	15		
INTERFACE	522	130	0		
TOTAL DEMANDS (D)	665	175	75		
TOTAL # OF DEMAND FAILURES (N_D)	14	5	2		
FAILURE PROBABILITY (N_D/D)	2.1×10^{-2}	2.86×10^{-2}	2.7×10^{-2}		

ASSUMPTIONS USED IN DETERMINING DEMANDS : Tests: MT and MG
1 test every 2 years: 1/2 × 10 yrs = 5 tests total × Population;
RV 1 test every 6 mos = 2 tests/1 yr × 10 yrs = 20 tests × population.
Auto & Manual: MT and MG = none assumed;
Assumed RV responds to 1 auto-start signal/year ×
10 years × Population

Figure 6.6 *Example Valve Demand/Failure Rate Worksheet. From Science Applications International Corporation.*

INTERFACING DEMANDS

ASSUMPTIONS

Assumed the following valve types are NOT used as isolation valves; Air Operated Diaphram, Solenoid Operated, Pressure Relief & Safety, Swing Check, or Tilting Disk Check.

ESTIMATION

From an inspection of the system P&ID it was determined that for a specific maintenance action;

a. 2 valve(s) is/are required to isolate each mT valve
(valve type)

b. 1 valve(s) is/are required to isolate each mG valve
(valve type)

c. 1 valve(s) is/are required to isolate each RV valve
(valve type)

d. ______ valve(s) is/are required to isolate each ______ valve
(valve type)

e. ______ valve(s) is/are required to isolate each ______ valve
(valve type)

f. ______ valve(s) is/are required to isolate each ______ valve
(valve type)

g. ______ valve(s) is/are required to isolate each ______ valve
(valve type)

h. ______ valve(s) is/are required to isolate each ______ valve
(valve type)

i. 3 valve(s) is/are required to isolate each PCP PUMP

j. ______ valve(s) is/are required to isolate each ______ HX

[EXAMPLE]
Total Interface Demands (I_{DT}) = (a x AO failures) + (b x MO failures) + (c x HW failures) + ... (j x Pump failures)

$(2 \times 43) + (1 \times 20) + (1 \times 15) + (3 \times 177) = 86 + 20 + 15 + 531$
$= 652$

I_{DT} = 652

POPULATION CALCULATION

Population of	mT	= 20
Population of	mG	= 5
Population of		=
Population of		=
Population of		=
Total Population (I_P)		= 25

mT: $20/25 = .8$

$.8 \times 652 = 522$

mG: $5/25 = .2$

$.2 \times 652 = 130$

Figure 6.7 *Example Interfacing Demands Sheet. From Science Applications International Corporation.*

PUMP DEMAND DATA SHEET

SYSTEM : PRIMARY COOLING #37

PUMP TYPE : PRIMARY COOLANT (PCP) DRIVER TYPE : TURBINE

TOTAL POPULATION : 6

PUMP ID / DEMANDS	PCP 1	PCP 2	PCP 3	PCP 4	PCP 5	PCP 6
TEST	X	X	X	X	X	X
AUTO. & MAN.	212	212	212	212	212	212
FAILURE RELATED	22	19	13	26	32	25
INTERFACE	X	X	X	X	X	X
TOTAL (D)	234	231	225	238	244	237
TOTAL # OF DEMAND FAILURES	1	0	1	0	0	0
FAILURE PROBABILITY	4.27×10^{-3}	1.43×10^{-3}	4.44×10^{-3}	1.39×10^{-3}	1.35×10^{-3}	1.39×10^{-3}

ASSUMPTIONS USED IN DETERMINING DEMANDS : The system is always in operation; there is no system testing. Assumed 1 demand to PCPs per each start-up of reactor. 5 out of 6 PCPs run continually, therefore, the # of starts per PCP = 5/6 × 254 (total # of reactor start-ups) = 212

Assumed that PCPs run an equal amount of time

Figure 6.8 *Example Pump Demand/Failure Rate Worksheet. From Science Applications International Corporation.*

PUMP FAILURE SUMMARY

PLANT ID NRx SYSTEM PRIMARY COOLING #37

PUMP ID	POP.	# of Demand Failures	# of Time-Related Failures	# of Demands	Exposure Time	Failure Probability	Failure Rate
PCP#1	1	1	4	234	64174	4.27×10^{-3}	6.23×10^{-5}
PCP#2	1	0	1	231	64174	1.43×10^{-3}	1.56×10^{-5}
PCP#3	1	1	2	225	64174	4.44×10^{-3}	3.12×10^{-5}
PCP#4	1	0	6	238	64174	1.39×10^{-3}	9.35×10^{-5}
PCP#5	1	0	2	244	64174	1.35×10^{-3}	3.12×10^{-5}
PCP#6	1	0	2	237	64174	1.39×10^{-3}	3.12×10^{-5}

Figure 6.9 *Example Pump Failure Rate Summary. From Science Applications International Corporation.*

DATA ON SELECTED PROCESS SYSTEMS AND EQUIPMENT	
Taxonomy No. 3.2.1.2	**Equipment Description** PIPING SYSTEMS - METAL-FITTINGS
Operating Mode	**Process Severity** UNKNOWN

Population	Samples	Aggregated time in service (10^6 hrs)		No. of Demands
		Calendar time	Operating time	

Failure mode	Failures (per 10^6 hrs)			Failures (per 10^3 demands)		
	Lower	Mean	Upper	Lower	Mean	Upper
CATASTROPHIC						
a. 0-10% Flow Area	9.57	551.0	2130.0			
b. >10% Flow Area	3.70	213.0	824.0			
c. Rupture						
d. Plugged						
DEGRADED						
a. Restricted Flow						
INCIPIENT						
a. Wall Thinning						
b. Embrittlement						
c. Cracked or Flawed						
d. Erratic Flow						

NOT TO BE USED

Equipment Boundary

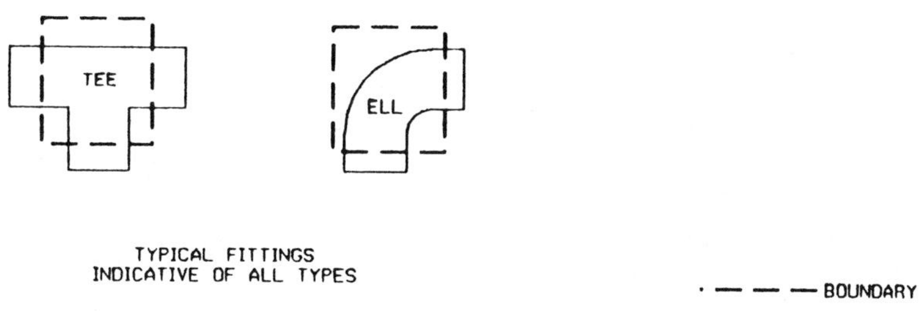

Data Reference No. (Table 5.1): 9

Figure 6.10. *Example of false failure rate data.*

6.5 Statistical Treatment

For demand-related failures, confidence intervals can be calculated from approximations using the Fisher (F), chi-square, normal, lognormal, or Weibull distribution, depending on the relationship between N (number of failures) and D (number of demands). The F distribution approximation is conservative but applies under all conditions. The chi-square approximation applies for large values of D, small values of N, and when $D/N>10$. The remaining approximations are used as appropriate when both N and D are large and both N and D/N are larger than 10. For references to these confidence intervals, see Nelson.[6] For time-related failures, estimation and confidence intervals are specific to the distribution that is selected for the time-dependent failure model. Widely used distributions include the exponential, normal, lognormal, Weibull, and gamma distributions. References such as Nelson[6,7] and Henley and Kumamoto[8] provide a comprehensive treatment of parameter estimation, reliability, failure rates, and uncertainties associated with these quantities. When calculating confidence intervals for time-dependent data, it is important to account for possible inaccuracies and incomplete data. The relative level of accuracy and completeness depends upon the ability to detect all failures, rounding errors in the collection process, and what data are actually collected.

When the underlying distribution is not known, tools such as histograms, probability curves, piecewise polynomial approximations, and general techniques are available to fit distributions to data. It may be necessary to assume an appropriate distribution in order to obtain the relevant parameters. Any assumptions made should be supported by manufacturer's data or data from the literature on similar items working in similar environments. Experience indicates that some probability distributions are more appropriate in certain situations than others. What follows is a brief overview on their applications in different environments. A more rigorous discussion of the statistics involved is provided in the CPQRA Guidelines.[9]

Exponential distribution: The exponential distribution is the simplest component life distribution. It is suited to model chance failures when the rate at which events occur remains constant over time. It is often suitable for the time between failures for repairable equipment.

Normal distribution: The normal distribution is the best known symmetric distribution, and two parameters completely describe the distribution. It often describes dimensions of parts made by automatic processes, natural and physical phenomena, and equipment that has increasing failure rates with time.

Lognormal distribution: Similar to a normal distribution. However, the logarithms of the values of the random variables are normally distributed. Typical applications are metal fatigue, electrical insulation life, time-to-repair data, continuous process (i.e., chemical processes) failure and repair data.

Weibull distribution: This distribution has been useful in a variety of reliability applications. The Weibull distribution is described by three parameters, and it can assume many shapes depending upon the values of the parameters. It can be used to model decreasing, increasing, and constant failure rates.

There are other distributions that can be used in a variety of reliability models. The Poisson, the extreme value, gamma, binomial, and Rayleigh distributions are sometimes used in specialized models.

References

1. Drago, J. P., Borkowski, R. J., Pike, D. H., and Goldberg, F. F. *The In-Plant Reliability Data Base for Nuclear Power Plant Components: Data Collection and Methodology Report*. NUREG/CR-2641, ORNL/TM-9216, January 1985.
2. Smith, D. J., and Babb, A. H. *Maintainability Engineering*. Pitman, London, 1973.
3. Bendell, A., "An Overview of Collection, Analysis and Application of Reliability Data in the Process Industries," *IEEE Transactions on Reliability*, Vol. 37, No. 2, June 1988, pp 132-137.
4. *EuReDatA Project No. 3, Guide to Reliability Data Collection and Management* B. Stevens (Ed.), Commission of the European Communities, Joint Research Centre, Ispra Establishment, S.P./1.05.E3.86.20.
5. *Guide to the Collection and Representation of Electrical, Electronic, Sensing Component, and Mechanical Equipment Reliability Data for Nuclear Generating Stations*. IEEE Std. 500 1984, Institute of Electrical and Electronic Engineers, New York, 1984.
6. Nelson, W.B. *Applied Life Data Analysis* John Wiley & Sons, New York 1982.
7. Nelson, W.B., "Basic Concepts and Distributions for Product Life," General Electric Technical Information Series, Report No. 74CRD311, 1974.
8. Henley, E.J. and Kumamoto, H. *Reliability Engineering and Risk Assessment*. Prentice-Hall, Englewood Cliffs, NJ, 1981.
9. Guidelines for Chemical Process Quantitative Risk Analysis. AIChE-CCPS, New York, 1989.

7
Failure Rate Data Transfer

The CCPS Taxonomy developed for this Guidelines and the plant data collection and conversion procedures discussed in Chapter 6 are a step toward the generation of a greater amount and more consistent failure rate data. By using accepted standard definitions and procedures for data collection and conversion, plant-specific failure rate data can be added to and combined with the generic data in this book or with other types of failure rate data discussed in the CPQRA Guidelines. If these additional failure rate data are made available to CCPS, its taxonomy can be amended and upgraded where needed, and the quality of the data in its Generic Failure Rate Data Base will improve for all CPI users. To facilitate the transfer of high quality, plant-specific failure rate data, a Failure Rate Data Transfer Form, Figure 7.1, has been provided. This form can be used to standardize the transfer of failure rate data to CCPS or to others in the CPI.

Industry	______________________
Geographic Area	______________________
Plant Identification	______________________
System Identification	______________________
Physical Description	______________________ ______________________
CCPS Taxonomy Number	______________________
Operating Mode	Alternating (Circle One) Running Standby
Process Severity	1- Clean (Circle One) 2- General Industry 3- Moderately Severe 4- Severe
Process Medium	______________________
Physical State	Vapor (Circle all that apply) Liquid Solid
Failure Mode	______________________
Population	______________________
Failure Type	Demand Related Time-Related (Circle One)
Number of Demands	______________________
Number of Failures	______________________
Operating Time	______________________hours
Calendar Time	Start date__________ End date__________

Figure 7.1 *Example Failure Rate Data Transfer Form.*

8
Supplemental References

The following references describe data collection, analysis and application techniques but in general do not contain reliability data:

Aird, R. J. "Practical Estimation of Reliability Parameters for Process Equipment." *Reliability Engineering,* Vol. 7, 1984, pp. 314–318.

American National Standard Nuclear Plant Reliability Data Collection and Reporting System. American Nuclear Society, Hinsdale, Illinois 60521, 1976.

Anderson, M. and K. Ekberg. "The ATV Data System and Its Use." Paper presented at the International ANS/ENS Topical Meeting on Probabilistic Safety Methods and Applications, San Francisco, February 1988.

Apostolakis, G. "Data Analysis in Risk Assessments." *Nuclear Engineering and Design,* Vol. 71, July 1982.

Apostolakis, G., S. Kaplan, B. J. Garrick, and R. J. Duphily. "Data Specialization for Plant Specific Risk Studies." *Nuclear Engineering and Design,* Vol. 56, 1980.

Atre, S. *Data Base: Structured Techniques for Design, Performance, and Management.* John Wiley & Sons, New York, 1980.

Bello, G. and A. Bobbio. "A Reliability Data Bank in the Petrochemical Sector." *Terotechnia,* Vol. 2, 1981, pp. 167–174.

Bendell, A. "Proportional Hazards Modeling in Reliability Assessment." *Reliability Engineering,* Vol. 2, 1983, pp. 175–183.

Billington, R. and R. N. Allan. *Reliability Evaluation of Engineering Systems: Concepts and Techniques,* Plenum Press, New York, 1983.

Blanks, H. S. "The Generation and Use of Component Failure Rate Data." *Quality Assurance,* Vol. 3, No. 3, September 1977, p. 85.

Bloch, H. P. *Improving Machinery Reliability.* Gulf Publishing, Houston, 1982.

Bockholts, P. "Collection and Application of Incident Data." *I. Chem. E. Symp.,* Series 80, K11–K21, 1983.

Bompass-Smith, J. H. *Mechanical Survival.* McGraw-Hill, New York, 1973.

Bucaram, S. M. and B. J. Yeary. "Data Gathering System to Optimize Production Operations: A 14-Year Overview." *J. Pet. Technol.,* Vol. 39, No. 4, April 1987, pp. 457–462.

Capobianci, S. "The Problem of Data Homogenization in Reliability Data Banks: A Scheme of Classifications." Paper 11.B.5, ANS/ENS Topical Meeting on PRA, September 1981.

Colombo, A. G. and R. J. Jaarsma. "Combination of Reliability Parameters from Different Data Sources." *Proceedings of the 4th EuReDatA Conference,* 1983.

Colombo, A. G. and O. Saracco. "Bayesian Estimation of the Time-independent Failure Rate of an Item Taking into Account Its Quality and Operational Constraints." *Proceedings of the 4th Euredata Conference,* 1983.

Common Cause Fault Rates for Pumps, U. S. Nuclear Regulatory Commission, NUREG/CR-2098, February 1983.

Common Cause Fault Rates for Valves, U. S. Nuclear Regulatory Commission, NUREG/CR-2770, February 1983.

Common Cause Fault Rates for Diesel Generators, U. S. Nuclear Regulatory Commission, NUREG/CR-2099, February 1983.

Cran, G. W. "Graphical Estimation Methods for Weibull Distribution." *Microelectronics and Reliability,* Vol. 15, 1976, p. 47.

Dhillon, B. S. and S. N. Rayapati. "Chemical System Reliability: A Review." *IEEE Transactions on Reliability,* Vol. 37, No. 2, June 1988, p. 199.

Eames, A.R. *Nuclear Engineering,* 11(118) March, 1966.

FACTS Incident Databank. TNO, Department of Industrial Safety, P.O. Box 342, 7300 Apeldoorn, The Netherlands.

Failure Data and Failure Analysis. In *Power and Processing Industries.* PVP-PB-023, ASME-G00123, American Society of Mechanical Engineers, New York, 1977.

Fienemann, M. and C. Mason. "Nuclear Plant Component Nomenclature Hierarchy for Failure and Repair Reporting." SAI/NY-R-81-4, 1981.

Fleming, K. N. *et al., Classification and Analysis of Reactor Operating Experience Involving Dependent Events.* EPRI NP-3967, Electric Power Research Institute, Palo Alto, California, 1985.

Fong, J. T. *et al., Inservice Data Reporting and Analysis.* American Society of Mechanical Engineers, New York, 1978.

Green, A.E., U.K. Atomic Energy Authority, Health and Safety Branch, Rep. AHSB (S) R113, Risley, Lancashire, 1966.

Green, A.E., and A.J. Bourne. *Reliability Technology.* John Wiley & Sons, New York, 1977.

Green, A.E., and A.J. Bourne, U.K. Atomic Energy Authority, Health and Safety Branch, Rep. AHSB (S) R117, Parts 1–3, Risley, Lancashire, 1966.

Green, A.E., and A.J. Bourne, U.K. Atomic Energy Authority, Health and Safety Branch, Rep. AHSB (S) R91, Risley, Lancashire, 1965.

Henley, E. J. and H. Kumamoto. *Reliability Engineering and Risk Assessment.* Prentice-Hall, Englewood Cliffs, New Jersey, 1981.

Hensley, G. *Measurement Control,* 1, T72, 1968.

Hensley, G., U.K. Atomic Energy Authority, Health and Safety Branch, Rep. AHSB (S) R136, Risley, Lancashire, 1967.

Hensley, G., U.K. Atomic Energy Authority, Health and Safety Branch, Rep. AHSB (S) R178, Risley, Lancashire, 1967.

Hensley, G., U.K. Atomic Energy Authority, Systems Reliability Service, Rep. SRS/G1/1, Culcheth, Lancashire.

Hester, O. V., S. R. Brown, and C. D. Gentillon. *Annotated Bibliography of Reliability and Risk Data Sources,* EGG-REQ-7827, Idaho National Engineering Laboratory, 1987.

Hoaglin, D. *et al., Understanding Robust and Exploratory Data Analysis.* John Wiley & Sons, New York, 1980.

Hoemke, P. "Experience in Germany with Reliability Data Assessment, Results and Current Problems." Paper II.B.2, ANS/ENS Topical Meeting on PRA, September, 1981.

Interim Reliability Evaluation Program (IREP) Procedures Guide, NUREG/CR-2728.

Kaplan, S. "On the Use of Data and Judgment in Probabilistic Risk and Safety Analysis." Paper presented at the Nuclear Engineering and Design International Post-Conference Seminar, August 26–27, SMiRT 8, Brussels, Belgium, 1985.

Kazarians, M. and R. F. Boykin. "Quantitative Risk Assessment for Chemical Operations." *Proceedings of the International Symposium on Preventing Major Chemical Accidents,* February 3–5, 1987, Washington, D.C. AIChE-CCPS, New York.

Kido, C. *A Bibliography of Data Bases for Nuclear Power Plant Risk Assessment.* EGG-EA-6100, November 1982.

Lees, F. P. *Loss Prevention in the Process Industries.* 2 Vols. Butterworths, Stoneham, Massachusetts 1980.

Leverenz, F. L. and D. C. Cox. *Probabilistic Risk Assessment Course Documentation,* Volume 6, NUREG/CR-4350, U. S. Nuclear Regulatory Commission, Washington, DC, 1985.

Libberton, G. P., A. Bendell, L. A. Walls, and A. G. Cannon. "Some Problems Associated with the Collection of Data for Automatic Fire Detection Systems on a Large Industrial Site." *Proc. I Mech. E. Conference on Data Collection and Analysis for Reliability Assessment.* London, England, 1986.

Mancini, G. "The European Reliability Data System (ERDS) State of the Art and Future Developments." Paper 11.B.1, ANS/ENS Topical Meeting on PRA, September 1981.

Mann, H., *et al. Methods for Statistical Analysis of Reliability and Life Data.* John Wiley & Sons, New York, 1974.

Mann, N. R., R. E. Schafer, and N. D. Singpurwalla. *Methods for Statistical Analysis of Reliability and Life Data. John Wiley & Sons, New York, 1983.*

Martz, H. F. and R. A. Waller. An Exploratory Comparison of Methods for Combining Failure Rate Data from Different Data Sources. Report No. LA-7556-MS, Los Alamos Scientific Laboratory, 1987.

Martz, H. F. and R. A. Waller. *Bayesian Reliability Analysis.* John Wiley & Sons, New York, 1982.

McCormick, N. J. *Reliability and Risk Analysis: Methods and Nuclear Power Applications.* Academic Press, San Diego, 1981.

Miller, M.J. "Reliability of Fire Protection Systems." *Chemical Engineering Progress,* Vol. 70, No. 4, April 1974, pp. 22–67.

Miller, M. J. "Risk Management and Reliability—An Update." Paper presented at the Conference on Fire Safety Evaluation, National Academy of Sciences, Washington, D.C., September 1978.

Moieni, P., G. Apostolakis, and G. E. Cummings. "On Random and Systematic Failures." *Reliability Engineering,* Vol. 2, No. 3, September 1981.

Mosleh, A. and G. Apostolakis. "Some Properties of Distributions Useful in the Study of Rare Events." *IEEE Transactions on Reliability,* Vol. R–31, April 1982, pp. 87–94.

Mosleh, A., M. Kazarians, and W. C. Gekler. "Development of a Risk and Reliability Data Base for the Chemical Industry." Paper presented at the American Institute of Chemical Engineers National Meeting, Boston, MA, 1984.

Mulvihill, R. J. and Y. G. Mody. "Failure Rate Distribution and Mean Time to Repair Estimates for Heat Exchanger and Condenser." SAN/RA50239, Pickard, Lowe and Garrick, Inc., November 1981.

National Reliability Evaluation Program (NREP) Procedures Guide, NUREG/CR-2815, 1982.

Olson, R. L., *et al. IEEE Guide for General Principles of Reliability Analysis of Nuclear Power Generating Station Safety Systems,* IEEE Std 352-1987. John Wiley & Sons, New York, 1987.

Oswald, A. J. Generic Data Base and Models Chapter of the National Reliability Evaluation Program (NREP) Guide. EG&G Idaho, Inc., EGG-EA-5887, June 1982.

Program Plan for an Integrated Risk Assessment Data Acquisition Program. U. S. Nuclear Regulatory Commission, September 1982.

Reporting Procedures Manual for the Nuclear Plant Reliability Data System. Southwest Research, San Antonio, Texas.

Shinha, S. K. and B. K. Kale *Life Testing and Reliability Estimation.* John Wiley & Sons (Halsted), New York, 1980.

Shooman, M. L. *Probabilistic Reliability: An Engineering Approach.* McGraw-Hill, New York, 1968.

Skala, V. *Instrument Technology,* 21(10), 27, 1974.

Smith, D. J. *Reliability and Maintainability in Perspective,* Macmillan, New York, 1985.

Smith, D. J. and A. H. Babb. *Maintainability Engineering,* Pitman, London, 1973.

Thomas, D. R. *et al.* "Inferences on the Parameters of the Weibull Distribution." *Technometrics,* Vol. II, 1969, pp. 445.

Thomas, H. M. "Pipe and Vessel Failure Probability." *Reliability Engineering,* Vol. 2, 1981, pp. 83–124.

Tukey, J. W., *Exploratory Data Analysis.* Addison-Wesley, Reading, Massachusetts, 1977.

Walls, L. A. and A. Bendell. "Time Series Methods in Reliability." *Proceedings of the 9th Advances in Reliability Technology Symposium.* Bradford, C2/3, 1986, pp. 1–18.

Weibull, W. "A Statistical Distribution Function of Wide Application." *J. Appl. Mech.,* Vol. 18, 1951, pp. 293.

Welker, E. L. and M. Lipow. "Estimating the Exponential Failure Rate from Data with No Failure Events." *Proceedings of the 1974 Annual Reliability and Maintainability Symposium.* Los Angeles, January 29–31, 1974 (IEEE Catalog No. 74CH08201ROC).

Wightman, D. W. and A. Bendell. "The Practical Application of Proportional Hazards Modeling." *Proceedings of the 5th National Reliability Conference.* Birmingham, England, 1985, 2B/3, pp. 1–16.

Appendix A
CCPS Generic Failure Rate Data Base Taxonomy

1.0 Electrical
2.0 Instrumentation
3.0 Process Equipment
4.0 Protection Systems
5.0 Utilities

Section 2.4 discusses the use of this CCPS Taxonomy.

1.0 ELECTRICAL EQUIPMENT
1.1 Motors

APPENDIX A - CCPS TAXONOMY

EQUIPMENT DESCRIPTION		SERVICE DESCRIPTION		FAILURE DESCRIPTION
		OPERATING MODE	PROCESS SEVERITY	
1.1.1 AC				1- Catastrophic
1.1.1.1 Induction				a. Fails to Run Once Started
1.1.1.1.1 Squirrel-cage	.1 0 - 250 V			b. Fails to Position Properly
	.2 >250 V - 600 V			c. Fails to Start on Demand
	.3 >600 V - 15 KV			2- Degraded
				a. Shaft Fails to Run at Rated Speed
				b. Fails to Position Correctly
1.1.1.1.2 Wound-Rotor	.1 0 - 600 V			c. "Hunts" for Correct Position
	.2 >600 V - 15 KV			
1.1.1.2 Synchronous				
1.1.1.2.1 Brushless	.1 0 - 600 V			
1.1.1.2.2 Reluctance (Clock Motors)	.2 >600 V - 5 KV			
1.1.1.2.3 Slip-Ring	.3 >5 KV - 8 KV			
	.4 >8 KV - 15 KV			
1.1.2 DC				
1.1.2.1 Commutator	.1 0 - 99 V			
1.1.2.2 Permanent-Magnet Field	.2 >99 V - 299 V			
	.3 >299 V			

APPENDIX A - CCPS TAXONOMY

EQUIPMENT DESCRIPTION		SERVICE DESCRIPTION		FAILURE DESCRIPTION
		OPERATING MODE	PROCESS SEVERITY	
1.2.1 Batteries/Battery Banks 1.2.1.1 <u>Lead Acid</u> 1.2.1.2 <u>Nickel Cadmium</u>	.1 0 - 99 V .2 >99 V - 299 V .3 >299 V			1- Catastrophic a. No Output Given Challenge 2- Degraded a. Low Output Given Challenge b. No Output During Test c. Low Output During Test 3- Incipient a. Leakage b. Physical Evidence of Damage or Deterioration
1.2.2 Battery Chargers	.1 0 - 99 V .2 >99 V - 299 V .3 >299 V			1- Catastrophic a. No electrical output 2- Degraded a. Low output b. High output c. Erratic output 3- Incipient a. Overheated

APPENDIX A - CCPS TAXONOMY

EQUIPMENT DESCRIPTION		SERVICE DESCRIPTION		FAILURE DESCRIPTION
		OPERATING MODE	PROCESS SEVERITY	
1.2.3 Circuit Breakers 1.2.3.1 AC 1.2.3.1.1 Molded-Case 1.2.3.1.1.1 Thermal - Magnetic Trips 1.2.3.1.1.2 Solid-State Trip	.1 0 - 219 V .2 >219 V - 440 V .3 >440 V			1- Catastrophic a. Spurious Operation b. Failed to Open on Demand c. Failed to Close on Demand 2- Degraded 3- Incipient a. Contaminated
1.2.3.1.2 Power 1.2.3.1.2.1 Air-Magnetic	.1 0 - 600 V .2 >600 V - 5 KV .3 >5 KV - 8 KV .4 >8 KV - 15 KV			
1.2.3.1.2.2 Oil 1.2.3.1.2.3 Vacuum 1.2.3.1.2.4 Solid State	.1 >600 V - 5 KV .2 >5 KV - 8 KV .3 >8 KV - 15 KV .4 >15 KV - 72.5 KV			
1.2.3.2 DC 1.2.3.2.1 Power 1.2.3.2.2 Molded Case 1.2.3.2.3 Solid State	.1 0 - 99 V .2 >99 V - 299 V .3 >299 V			

APPENDIX A - CCPS TAXONOMY

EQUIPMENT DESCRIPTION		SERVICE DESCRIPTION		FAILURE DESCRIPTION
		OPERATING MODE	PROCESS SEVERITY	
1.2.4 Inverters				1- Catastrophic a. No Output b. Fails to Transfer 2- Degraded a. Output frequency outside specification b. Output Voltage outside spec. c. Output current outside spec. 3- Incipient a. Overheating b. Faulty indication c. Dirt/dust Contamination
1.2.5 Electrochemical Processing Rectifiers (10 - 200 mw)				1- Catastrophic a. Short Circuit 2- Degraded a. Diodes short out b. Thyristors short out
1.2.6 Fuses 1.2.6.1 Control	.1 0 - 120 V .2 >120 V			All Modes 1- Catastrophic a. Premature Open
1.2.6.2 Power	.1 0 - 250 V .2 >250 V - 600 V .3 >600 V - 5 KV .4 >5 KV - 8 KV .5 >8 KV - 15 KV			

1.0 **ELECTRICAL EQUIPMENT**
1.2 **Power Conditioning and Protection Devices**

APPENDIX A - CCPS TAXONOMY

EQUIPMENT DESCRIPTION			SERVICE DESCRIPTION		FAILURE DESCRIPTION
			OPERATING MODE	PROCESS SEVERITY	
1.2.7 Relays **1.2.7.1** <u>Control</u> **1.2.7.2** <u>Protective</u> **1.2.7.3** <u>Time Delay</u>	.1 Electromagnetic .2 Solid-State .3 Electromag/solid state hybrid	.1 0 - 2 break amps .2 >2 - 5 break amps .3 >5 - 10 break amps	.1 Energized, contacts closed .2 Energized, contacts open .3 De-energized, contacts closed .4 De-energized, contacts open		1- Catastrophic a. Fails to Close on Demand b. Fails to Open on Demand c. Spurious Operation d. Delayed Change of State e. Premature Change of State
1.2.8 Transformers **1.2.8.1** <u>Power</u> **1.2.8.2** <u>Instrument/Control</u> **1.2.8.3** <u>Rectifier</u>	.1 Liquid .2 Dry				All modes 1- Catastrophic a. No Output 2- Degraded a. Incorrect Output 3- Incipient a. Insulation Integrity
1.2.9 Uninterruptible Power Supplies (Rectifer + Solid State Inverter + Battery)					1- Catastrophic a. No output b. Output Outside Specification 2- Degraded a. Output slightly outside spec. 3- Incipient a. Leakage b. Smoking c. Noisy

1.0 ELECTRICAL EQUIPMENT
1.3 Power Generation

APPENDIX A - CCPS TAXONOMY

EQUIPMENT DESCRIPTION		SERVICE DESCRIPTION		FAILURE DESCRIPTION
		OPERATING MODE	PROCESS SEVERITY	
1.3.1 Emergency Power Generators 1.3.1.1 Diesel Driven 1.3.1.2 Gas Driven				1- Catastrophic a. Fails to Start on Demand b. Fails to Run once Started c. Fails to Supply Load in Time 2- Degraded a. Delayed Start after Multiple Attempts b. Fails to Maintain Voltage/ Frequency Spec. 3- Incipient a. Improper Cooling/Heating b. RPM Hunting c. Faulty Indication d. Vibration e. Improper Lubrication
1.3.2 Main Power Generators 1.3.2.1 Diesel Engine Driven 1.3.2.1.1 Turbo Charged 1.3.2.1.2 Non-Turbo Charged 1.3.2.2 Gasoline Engine Driven (50 - 100 KW) 1.3.2.3 Reciprocating Engines - Natural Gas Powered 1.3.2.4 Turbine Driven 1.3.2.4.1 Gas 1.3.2.4.2 Industrial 1.3.2.4.2.1 Dual Fuel 1.3.2.4.2.2 Single Fuel 1.3.2.4.3 Aero derivative	.1 0 - 1000 KW .2 >1000 KW	Continuous Standby		1- Catastrophic a. Fails to Run Once Started b. Fails to Position Properly c. Fails to Start on Demand 2- Degraded a. Shaft Fails to Run @ Rated Speed b. Fails to Position Properly c. "Hunts" for Correct Position 3- Incipient a. Overheating b. Vibration

2.0 INSTRUMENTATION
2.1 Process Wetted & Field Instrumentation

APPENDIX A - CCPS TAXONOMY

EQUIPMENT DESCRIPTION	SERVICE DESCRIPTION		FAILURE DESCRIPTION
	OPERATING MODE	PROCESS SEVERITY	
2.1.1 Analyzers 2.1.1.1 Electromagnetic Radiation Interaction 2.1.1.1.1 Optical Emission Spectrochemical 2.1.1.1.2 Flame Photometric 2.1.1.1.3 Fluorescence 2.1.1.1.4 Ultraviolet-Absorption 2.1.1.1.5 Colorimeter 2.1.1.1.6 Nondispersive Infra Red Analyzer 2.1.1.2 Chemical Interaction 2.1.1.2.1 Orsat 2.1.1.2.2 Titrator, Automatic 2.1.1.2.3 Impregnated Paper Tape 2.1.1.2.4 Luminescence Meter 2.1.1.2.5 Combustible (catalytic Detector) 2.1.1.2.6 Redox Potentiometry (ORP) 2.1.1.2.7 pH 2.1.1.2.7.1 Chemical Indicators 2.1.1.2.7.2 Potentiometric Meters 2.1.1.2.8 Metal Ion Equilibria 2.1.1.2.9 Calorimeter 2.1.1.3 Electric/Magnetic Field Reaction 2.1.1.3.1 Mass Spectrometer 2.1.1.3.2 Electrical Conductivity 2.1.1.3.3 Paramagnetic 2.1.1.3.4 Electrochemical 2.1.1.3.4.1 Dry Cell 2.1.1.3.4.2 Wet Cell 2.1.1.3.5 Nuclear Magnetic Resonance 2.1.1.3.6 Electron Paramagnetic Resonance 2.1.1.4 Thermal/Mechanical Energy Interaction 2.1.1.4.1 Thermal Conductivity 2.1.1.4.2 Hygrometer 2.1.1.4.3 Vapor Equilibrium Dewpoint Meter 2.1.1.4.4 Chromatograph 2.1.1.4.5 Viscometer 2.1.1.4.6 Densitometer			1- Catastrophic a. Maximum Output b. No Output c. No change of Output with Change of Input 2- Degraded a. Erratic Output b. High Output c. Low Output 3- Incipient

APPENDIX A - CCPS TAXONOMY

EQUIPMENT DESCRIPTION	SERVICE DESCRIPTION		FAILURE DESCRIPTION
	OPERATING MODE	PROCESS SEVERITY	
2.1.2 Controllers **2.1.2.1** Flow **2.1.2.2** Level **2.1.2.3** Pressure **2.1.2.4** Temperature **2.1.2.5** Speed		1-4	1- Catastrophic a. Maximum Output b. No Output c. No Change of Output with Change of Input 2- Degraded a. Erratic Output b. High Output c. Low Output 3- Incipient

2.0 INSTRUMENTATION
2.1 Process Wetted & Field Instrumentation

APPENDIX A - CCPS TAXONOMY

EQUIPMENT DESCRIPTION	SERVICE DESCRIPTION		FAILURE DESCRIPTION
	OPERATING MODE	PROCESS SEVERITY	
2.1.3 Transmitters 2.1.3.1 <u>Electronic</u> 2.1.3.1.1 Flow 2.1.3.1.1.1 Differential Pressure 2.1.3.1.1.2 Variable Area 2.1.3.1.1.3 Magmeter 2.1.3.1.1.4 Mass Flow 2.1.3.1.1.5 Vortex Shedding 2.1.3.1.1.6 Turbine 2.1.3.1.2 Level 2.1.3.1.2.1 Differential Pressure 2.1.3.1.2.2 Displacement 2.1.3.1.2.3 Capacitance Probe 2.1.3.1.2.4 Float 2.1.3.1.2.5 Nuclear 2.1.3.1.3 Pressure 2.1.3.1.3.1 Capacitive 2.1.3.1.3.2 Strain Gauge 2.1.3.1.4 Temperature 2.1.3.1.4.1 Thermocouple 2.1.3.1.4.2 Resistance Temperature 2.1.3.1.4.3 Capillary & Bulb 2.1.3.1.5 Differential Pressure 2.1.3.1.6 Vibration 2.1.3.1.7 Speed 2.1.3.1.8 Weight 2.1.3.1.8.1 Load Cell 2.1.3.2 <u>Pneumatic</u> 2.1.3.2.1 Flow 2.1.3.2.1.1 Differential Pressure 2.1.3.2.1.2 Variable Area 2.1.3.2.2 Level 2.1.3.2.2.1 Differential Pressure 2.1.3.2.2.2 Displacer 2.1.3.2.2.3 Float 2.1.3.2.3 Pressure 2.1.3.2.3.1 Torque Tube 2.1.3.2.4 Temperature (Capillary & Bulb) 2.1.3.2.5 Differential Pressure		1-4	1- Catastrophic a. Maximum Output b. No Output c. No Change of Output with Change of Input 2- Degraded a. Erratic Output b. High Output c. Low Output 3- Incipient

APPENDIX A - CCPS TAXONOMY

EQUIPMENT DESCRIPTION	SERVICE DESCRIPTION		FAILURE DESCRIPTION
	OPERATING MODE	PROCESS SEVERITY	
2.1.4 Switches 2.1.4.1 Electric 2.1.4.1.1 Flow 2.1.4.1.1.1 Differential Pressure 2.1.4.1.1.2 Paddle 2.1.4.1.1.3 Thermal 2.1.4.1.2 Level 2.1.4.1.2.1 Differential Pressure 2.1.4.1.2.2 Displacement 2.1.4.1.2.3 Float 2.1.4.1.2.4 Capacitance Probe 2.1.4.1.3 Pressure 2.1.4.1.3.1 Bellows 2.1.4.1.3.2 Diaphragm 2.1.4.1.3.3 Bourdon Tube 2.1.4.1.3.4 Piston 2.1.4.1.4 Temperature 2.1.4.1.4.1 Bimetallic 2.1.4.1.4.2 Vapor Pressure Bulb 2.1.4.1.5 Differential Pressure 2.1.4.1.6 Vibration 2.1.4.1.7 Speed 2.1.4.2 Pneumatic 2.1.4.2.1 Flow 2.1.4.2.2 Level 2.1.4.2.3 Pressure 2.1.4.2.4 Temperature		1-4	1- Catastrophic a. Functioned without Signal b. Failed to Function when Signaled 2- Degraded a. Functioned at Improper Signal Level b. Intermittent Operation 3- Incipient a. In-service Problems
2.1.5 Flame Detectors 2.1.5.1 IR 2.1.5.2 Thermal Sensor 2.1.5.3 UV			1- Catastrophic a. Functioned without Signal b. Failed to Function when Signaled 2- Degraded a. Functioned at Improper Signal Level b. Intermittent Operation 3- Incipient a. In-service Problems

2.0 INSTRUMENTATION
2.1 Process Wetted & Field Instrumentation

APPENDIX A - CCPS TAXONOMY

EQUIPMENT DESCRIPTION	SERVICE DESCRIPTION		FAILURE DESCRIPTION
	OPERATING MODE	PROCESS SEVERITY	
2.1.6 Indicators/Element **2.1.6.1** Flow **2.1.6.1.1** Variable Area **2.1.6.1.2** Turbine **2.1.6.1.3** Positive Displacement **2.1.6.2** Level **2.1.6.2.1** Float **2.1.6.2.2** Displacer **2.1.6.2.3** Armored Sight Glass **2.1.6.2.4** Glass Tube Sight Glass **2.1.6.2.5** Differential Pressure **2.1.6.3** Pressure **2.1.6.3.1** Bourdon Tube **2.1.6.3.2** Helical Coil **2.1.6.4** Temperature **2.1.6.4.1** Thermocouple **2.1.6.4.2** Resistance Temperature Detector (RTD) **2.1.6.4.3** Dial **2.1.6.4.4** Vapor Pressure Bulb **2.1.6.4.5** Optical Pyrometer **2.1.6.4.6** Radiation Pyrometer		1-4	1- Catastrophic a. Maximum Output b. No Output c. No Change ofOoutput with Change of Input 2- Degraded a. Erratic Output b. High Output c. Low Output 3- Incipient
2.1.7 Recorders (Large Case) **2.1.7.1** Pressure **2.1.7.2** Temperature		1-4	1- Catastrophic a. Maximum Output b. No Output c. No change of Position with Change of Input 2- Degraded a. Erratic Output b. High Output c. Low Output 3- Incipient

2.0 **INSTRUMENTATION**
2.1 **Process Wetted & Field Instrumentation**

APPENDIX A - CCPS TAXONOMY

EQUIPMENT DESCRIPTION	SERVICE DESCRIPTION		FAILURE DESCRIPTION
	OPERATING MODE	PROCESS SEVERITY	
2.1.8 Transducers 2.1.8.1 Current to Pneumatic 2.1.8.2 Pneumatic to Current			1- Catastrophic a. Maximum Output b. No Output c. No Change of Output with Change of Input 2- Degraded a. Erratic Output b. High Output c. Low Output 3- Incipient

2.0 **INSTRUMENTATION**
2.2 **Control Room Instrumentation**

APPENDIX A - CCPS TAXONOMY

EQUIPMENT DESCRIPTION	SERVICE DESCRIPTION		FAILURE DESCRIPTION
	OPERATING MODE	PROCESS SEVERITY	
2.2.1 Controllers 2.2.1.1 <u>Electronic Panelboard (single loop)</u> 2.2.1.2 <u>Pneumatic Panelboard (single loop)</u>			1- Catastrophic a. Maximum Output b. No Output c. No Change of Output with Change of Input 2- Degraded a. Erratic Output b. High Output c. Low Output 3- Incipient
2.2.2 Annunciators			1- Catastrophic a. Operates Spuriously b. Fails to Operate on Demand 2- Degraded a. Operates at Low Intensity 3- Incipient a. Arcing
2.2.3 Indicators 2.2.3.1 <u>Electronic (single loop)</u> 2.2.3.2 <u>Pneumatic</u>			1- Catastrophic a. Maximum Output b. No Output c. Stuck in Position 2- Degraded a. Erratic Output b. High Output c. Low Output 3- Incipient
2.2.4 Recorders 2.2.4.1 <u>Electronic</u> 2.2.4.2 <u>Pneumatic</u>			1- Catastrophic a. Maximum Output b. No Output c. No change of position with change of input 2- Degraded a. Erratic Output b. High Output c. Low Output 3- Incipient

APPENDIX A - CCPS TAXONOMY

EQUIPMENT DESCRIPTION	SERVICE DESCRIPTION		FAILURE DESCRIPTION
	OPERATING MODE	PROCESS SEVERITY	
2.2.5 Computational Modules (Pneumatic)			1- Catastrophic a. Maximum Output b. No Output c. No Change of Output with Change of Input 2- Degraded a. Erratic Output b. High Output c. Low Output
2.2.6 Totalizers (Pneumatic)			1- Catastrophic a. No Output 2- Degraded a. Erratic Output b. High Output c. Low Output
2.2.7 Power Supplies (110 VAC Supply, 24 - 48 VDC Output)			1- Catastrophic a. No Output 2- Degraded a. Erratic Output b. High Output c. Low Output
2.2.8 Weight Scales			1- Catastrophic a. Maximum Output b. No Output c. No Change of Output 2- Degraded a. High Output b. Low Output

APPENDIX A - CCPS TAXONOMY

EQUIPMENT DESCRIPTION	SERVICE DESCRIPTION		FAILURE DESCRIPTION
	OPERATING MODE	PROCESS SEVERITY	
2.2.9 Switches (Electronic) 2.2.9.1 <u>Voltage (VDC) Input</u> 2.2.9.2 <u>Current Input</u> 2.2.9.3 <u>Theomocouple Input</u>			1- Catastrophic a. Functioned without Signal b. Failed to Function when Signaled 2- Degraded a. Functioned at Improper Signal Level b. Intermittent Operation 3- Incipient a. In-service Problems
2.2.10 Distributed Control Systems (Not Developed) **2.2.11 Programmable Controllers (Not Developed)**			

3.0 PROCESS EQUIPMENT
3.1 Heat Transfer Devices

APPENDIX A - CCPS TAXONOMY

EQUIPMENT DESCRIPTION	SERVICE DESCRIPTION		FAILURE DESCRIPTION
	OPERATING MODE	PROCESS SEVERITY	
3.1.1 Fired **3.1.1.1** Direct Contact **3.1.1.1.1** No Internals		1 - 4	1- Catastrophic a. Leakage 0 - 1/4" b. Leakage >1/4" c. Rupture 2- Degraded 3- Incipient a. Wall Thinning b. Embrittlement c. Cracked or Flawed
3.1.1.1.2 Static Internals **3.1.1.1.2.1** Tubed		1 - 4	1- Catastrophic a. 0 - 10% Flow Area b. >10% Flow Area c. Rupture d. Plugging 2- Degraded a. Restricted Flow 3- Incipient a. Wall Thinning b. Embrittlement c. Cracked or Flawed d. Erratic Flow
3.1.1.1.2.2 Non-Tubed		1 - 4	1- Catastrophic a. Leakage 0 - 1/4" b. Leakage >1/4" c. Rupture 2- Degraded 3- Incipient a. Wall Thinning b. Embrittlement c. Cracked or Flawed

APPENDIX A - CCPS TAXONOMY

EQUIPMENT DESCRIPTION	SERVICE DESCRIPTION		FAILURE DESCRIPTION
	OPERATING MODE	PROCESS SEVERITY	
3.1.1 Fired (Cont'd.) **3.1.1.1** Direct Contact (Cont'd.) **3.1.1.1.3** Moving Internals		1 - 4	1- Catastrophic a. Fails While Running b. Rupture (External Leakage) c. Internal Leakage d. Fails to Start on Demand e. Fails to Stop on Demand
3.1.1.2 Indirect Contact **3.1.1.2.1** No Internals		1 - 4	1- Catastrophic a. Leakage 0 - 1/4" b. Leakage >1/4" c. Rupture 2- Degraded 3- Incipient a. Wall Thinning b. Embrittlement c. Cracked or Flawed
3.1.1.2.2 Static Internals **3.1.1.2.2.1** Tubed		1 - 4	1- Catastrophic a. 0 - 10% Flow Area b. >10% Flow Area c. Rupture d. Plugging 2- Degraded a. Restricted Flow 3- Incipient a. Wall Thinning b. Embrittlement c. Cracked or Flawed d. Erratic Flow

APPENDIX A - CCPS TAXONOMY

EQUIPMENT DESCRIPTION	SERVICE DESCRIPTION		FAILURE DESCRIPTION
	OPERATING MODE	PROCESS SEVERITY	
3.1.1 Fired (Cont'd.) **3.1.1.2** Indirect Contact (Cont'd.) **3.1.1.2.2** Static Internals (Cont'd.) **3.1.1.2.2.2** Non-Tubed		1 - 4	1- Catastrophic a. Leakage 0 - 1/4" b. Leakage >1/4" c. Rupture 2- Degraded 3- Incipient a. Wall Thinning b. Embrittlement c. Cracked or Flawed
3.1.1.2.3 Moving Internals		1 - 4	1- Catastrophic a. Fails While Running b. Rupture (External Leakage) c. Internal Leakage d. Fails to Start on Demand e. Fails to Stop on Demand
3.1.2 Non-Fired **3.1.2.1** Direct Contact **3.1.2.1.1** No Internals		1 - 4	1- Catastrophic a. Leakage 0 - 1/4" b. Leakage >1/4" c. Rupture 2- Degraded 3- Incipient a. Wall Thinning b. Embrittlement c. Cracked or Flawed

3.0 PROCESS EQUIPMENT
3.1 Heat Transfer Devices

APPENDIX A - CCPS TAXONOMY

EQUIPMENT DESCRIPTION	SERVICE DESCRIPTION		FAILURE DESCRIPTION
	OPERATING MODE	PROCESS SEVERITY	
3.1.2 Non-Fired (Cont'd.) **3.1.2.1** Direct Contact (Cont'd.) **3.1.2.1.2** Static Internals **3.1.2.1.2.1** Tubed		1 - 4	1- Catastrophic a. 0 - 10% Flow Area b. >10% Flow Area c. Rupture d. Plugging 2- Degraded a. Restricted Flow 3- Incipient a. Wall Thinning b. Embrittlement c. Cracked or Flawed d. Erratic Flow
3.1.2.1.2.2 Non-Tubed		1 - 4	1- Catastrophic a. Leakage 0 - 1/4" b. Leakage >1/4" c. Rupture 2- Degraded 3- Incipient a. Wall Thinning b. Embrittlement c. Cracked or Flawed
3.1.2.1.3 Moving Internals		1 - 4	1- Catastrophic a. Fails While Running b. Rupture (External Leakage) c. Internal Leakage d. Fails to Start on Demand e. Fails to Stop on Demand

APPENDIX A - CCPS TAXONOMY

EQUIPMENT DESCRIPTION			SERVICE DESCRIPTION		FAILURE DESCRIPTION
			OPERATING MODE	PROCESS SEVERITY	
3.1.2 Non-Fired (Cont'd.) **3.1.2.2** <u>Indirect Contact</u> **3.1.2.2.1** Tubed **3.1.2.2.2** Non-Tubed	.1 Baffled .2 Non-Baffled	.1 Moving		1 - 4	1- Catastrophic a. Fails While Running b. Rupture (External Leakage) c. Internal Leakage d. Fails to Start on Demand e. Fails to Stop on Demand
		.2 Static		1 - 4	1- Catastrophic a. Leakage 0 - 1/4" b. Leakage >1/4" c. Rupture 2- Degraded 3- Incipient a. Wall Thinning b. Embrittlement c. Cracked or Flawed

3.0 PROCESS EQUIPMENT
3.2 Piping Systems

APPENDIX A - CCPS TAXONOMY

EQUIPMENT DESCRIPTION			SERVICE DESCRIPTION		FAILURE DESCRIPTION
			OPERATING MODE	PROCESS SEVERITY	
3.2.1 Metal				1 - 4	1- Catastrophic a. 0 - 10% Flow Area b. >10% Flow Area c. Rupture d. Plugged 2- Degraded a. Restricted Flow 3- Incipient a. Wall Thinning b. Embrittlement c. Cracked or Flawed d. Erratic Flow
3.2.1.1 <u>Straight Sections</u>	.1 1/4" - 1 1/2" .2 >1 1/2" - 4" .3 >4" - 6" .4 >6" - 10" .5 >10" - 16" .6 >16"	.1 Seamless .2 Welded			
3.2.1.2 <u>Fittings</u>					
3.2.1.3 <u>Penetrations</u>	.1 1/4" - 1 1/2" .2 >1 1/2" - 4" .3 >4" - 6" .4 >6" - 10" .5 >10" - 16" .6 >16"	.1 Thermowell .2 Threadolet .3 Weldolet			
3.2.1.4 <u>Connections</u>	.1 1/4" - 1 1/2" .2 >1 1/2" - 4" .3 >4" - 6" .4 >6" - 10" .5 >10" - 16" .6 >16"	.1 Brazed .2 Flanged .3 Threaded .4 Union .5 Welded			
3.2.1.5 <u>Welds</u>	.1 1/4" - 1 1/2" .2 >1 1/2" - 4" .3 >4" - 6" .4 >6" - 10" .5 >10" - 16" .6 >16"	.1 Factory .2 Field			

APPENDIX A - CCPS TAXONOMY

EQUIPMENT DESCRIPTION			SERVICE DESCRIPTION		FAILURE DESCRIPTION
			OPERATING MODE	PROCESS SEVERITY	
3.2.2 Lined				1 - 4	1- Catastrophic a. 0 - 10% Flow Area b. >10% Flow Area c. Rupture d. Plugged 2- Degraded a. Restricted Flow 3- Incipient a. Wall Thinning b. Embrittlement c. Cracked or Flawed d. Erratic Flow
3.2.2.1 Straight Sections	.1 0 - 6" .2 >6"	.1 Seamless .2 Welded			
3.2.2.2 Fittings	.1 0 - 6" .2 >6"				
3.2.2.3 Connections	.1 0 - 6" .2 >6"	.1 Brazed .2 Flanged .3 Threaded .4 Welded			
3.2.2.4 Penetrations					
3.2.2.5 Welds					
3.2.3 Rigid Plastic					
3.2.3.1 Straight Sections	.1 0 - 6" .2 >6"				
3.2.3.2 Fittings	.1 0 - 6" .2 >6"				
3.2.3.3 Connections	.1 0 - 6" .2 >6"	.1 Cemented .2 Threaded			
3.2.3.4 Penetrations (Cemented)					
3.2.4 Tubing Systems					
3.2.4.1 Tubing					
3.2.4.1.1 Metal					
3.2.4.1.2 Non-Metal					
3.2.4.2 Fittings					
3.2.4.2.1 Metal					
3.2.4.2.2 Non-Metal					
3.2.5 Hoses					
3.2.5.1 Metal	.1 0 - 4" .2 >4"				
3.2.5.2 Non-Metal					

APPENDIX A - CCPS TAXONOMY

EQUIPMENT DESCRIPTION			SERVICE DESCRIPTION		FAILURE DESCRIPTION
			OPERATING MODE	PROCESS SEVERITY	
3.3.1 **Centrifuges** (Driver Included) 3.3.1.1 Belt 3.3.1.2 Direct 3.3.1.3 Hydraulic	.1 Batch .2 Continuous	.1 Liquid-Liquid .2 Liquid-Solid	Alternating Running Standby	2 - 4	1- Catastrophic a. Fails while Running b. Rupture c. Spurious Start/Command Fault d. Fails to Start on Demand e. Fails to Stop on Demand 2- Degraded a. External Leakage

APPENDIX A - CCPS TAXONOMY

EQUIPMENT DESCRIPTION			SERVICE DESCRIPTION		FAILURE DESCRIPTION
			OPERATING MODE	PROCESS SEVERITY	
3.3.2 Compressors **3.3.2.1** <u>Electric Motor-Driven</u> **3.3.2.1.1** 0 - 200 psi	.1 Single Stage	.1 Centrifugal .2 Positive Displacement	Alternating Running Standby	1 - 3	1- Catastrophic a. Fails while running b. Rupture c. Spurious Start/Command fault d. Fails to Start on demand e. Fails to stop on demand 2- Degraded a. External Leakage
3.3.2.1.2 >200 - 3000 psi	.1 Single Stage	.1 Centrifugal .2 Positive Displacement	Standby Alternating Running	1 - 3	1- Catastrophic a. Fails while running b. Rupture c. Spurious Start/Command fault d. Fails to Start on demand e. Fails to stop on demand 2- Degraded a. External Leakage
	.2 Multiple Stage	.1 Centrifugal .2 Positive Displacement	Alternating Running	1 & 2	
3.3.2.1.3 >3000 psi	.1 Multiple Stage	.1 Centrifugal .2 Positive Displacement	Alternating Running	1 & 2	1- Catastrophic a. Fails while running b. Rupture c. Spurious Start/Command fault d. Fails to Start on demand e. Fails to stop on demand 2- Degraded a. External Leakage

APPENDIX A - CCPS TAXONOMY

EQUIPMENT DESCRIPTION		SERVICE DESCRIPTION		FAILURE DESCRIPTION
		OPERATING MODE	PROCESS SEVERITY	
3.3.2 Compressors (Cont'd.) **3.3.2.2** <u>Combustion Motor-Driven</u> **3.3.2.2.1** Diesel **3.3.2.2.1.1** 0 - 200 psi **3.3.2.2.1.1.1** Single Stage **3.3.2.2.1.1.1.1** Positive Displacement		Alternating Running Standby	1 - 3	1- Catastrophic a. Fails while running b. Rupture c. Spurious Start/Command fault d. Fails to Start on demand e. Fails to stop on demand 2- Degraded a. External Leakage
3.3.2.2.1.2 >200 - 3000 psi	.1 Single Stage .2 Multiple Stage	Alternating Running Standby Alternating Running	1 - 3 1 & 2	1- Catastrophic a. Fails while running b. Rupture c. Spurious Start/Command fault d. Fails to Start on demand e. Fails to stop on demand 2- Degraded a. External Leakage
3.3.2.2.2 Gas **3.3.2.2.2.1** 0 - 200 psi **3.3.2.2.2.1.1** Single Stage **3.3.2.2.2.1.1.1** Positive Displacement		Alternating Running Standby	1 - 3	1- Catastrophic a. Fails while running b. Rupture c. Spurious Start/Command fault d. Fails to Start on demand e. Fails to stop on demand 2- Degraded a. External Leakage

APPENDIX A - CCPS TAXONOMY

EQUIPMENT DESCRIPTION		SERVICE DESCRIPTION		FAILURE DESCRIPTION
		OPERATING MODE	PROCESS SEVERITY	
3.3.2 **Compressors** (Cont'd.) 3.3.2.2 Combustion Motor-Driven (Cont'd.) 3.3.2.2.2 Gas (Cont'd.) 3.3.2.2.2.2 >200 - 3000 psi	.1 Single Stage	Alternating Running Standby	1 - 3	1- Catastrophic a. Fails while running b. Rupture c. Spurious Start/Command fault d. Fails to Start on demand e. Fails to stop on demand 2- Degraded a. External Leakage
	.2 Multiple Stage	Alternating Running	1 & 2	
3.3.2.2.3 Gasoline 3.3.2.2.3.1 0 - 200 psi 3.3.2.2.3.1.1 Single Stage 3.3.2.2.3.1.1.1 Positive Displacement		Alternating Running Standby	1 - 3	1- Catastrophic a. Fails while running b. Rupture c. Spurious Start/Command fault d. Fails to Start on demand e. Fails to stop on demand 2- Degraded a. External Leakage
3.3.2.2.3.2 >200 - 3000 psi	.1 Single Stage	Alternating Running Standby	1 - 3	1- Catastrophic a. Fails while running b. Rupture c. Spurious Start/Command fault d. Fails to Start on demand e. Fails to stop on demand 2- Degraded a. External Leakage
	.2 Multiple Stage	Alternating Running	1 & 2	

APPENDIX A - CCPS TAXONOMY

EQUIPMENT DESCRIPTION			SERVICE DESCRIPTION		FAILURE DESCRIPTION
			OPERATING MODE	PROCESS SEVERITY	
3.3.2 Compressors (Cont'd.) **3.3.2.3** Turbine Driven **3.3.2.3.1** Gas **3.3.2.3.1.1** 0 - 200 psi **3.3.2.3.1.1.1** Single Stage		.1 Centrifugal .2 Positive Displacement	Alternating Running Standby	1 - 3	1- Catastrophic a. Fails while running b. Rupture c. Spurious Start/Command fault d. Fails to Start on demand e. Fails to stop on demand 2- Degraded a. External Leakage
3.3.2.3.2 >200 - 3000 psi	.1 Single Stage	.1 Centrifugal .2 Positive Displacement	Alternating Running Standby	1 - 3	1- Catastrophic a. Fails while running b. Rupture c. Spurious Start/Command fault d. Fails to Start on demand e. Fails to stop on demand 2- Degraded a. External Leakage
	.2 Multiple Stage	.1 Centrifugal .2 Positive Displacement	Alternating Running	1 & 2	

APPENDIX A - CCPS TAXONOMY

EQUIPMENT DESCRIPTION		SERVICE DESCRIPTION		FAILURE DESCRIPTION
		OPERATING MODE	PROCESS SEVERITY	
3.3.3 Blowers 3.3.3.1 <u>Motor-Driven</u>	.1 Direct .2 Gear	Alternating Running Standby	1 - 3	1- Catastrophic a. Fails While Running b. Influx of Contaminants (Backflow)
3.3.3.2 <u>Turbine-Driven</u>	.1 Gas .2 Steam .3 Direct .4 Gear	Running	1 - 3	
3.3.4 Motor Driven Fans	.1 Belt .2 Chain .3 Direct .4 Gear	Alternating Running Standby	1 & 2	1- Catastrophic a. Fails While Running b. Spurious Start/Command Fault (per standby hour) c. Fails to Start on Demand d. Fails to Stop on Demand
3.3.5 Extruders 3.3.5.1 <u>Heated</u> 3.3.5.2 <u>Non-Heated</u> 3.3.5.3 <u>Cooled</u> 3.3.5.4 <u>Non-Cooled</u>		Alternating Running	1 - 4	1- Catastrophic a. Fails while Running a.1 Motor Failure a.2 Screw Failure a.3 High Viscosity b. Seal Rupture - External c. Fails to Start on Demand d. Fails to Stop on Demand
3.3.6 Mixers/Blenders 3.3.6.1 <u>Mixers</u>	.1 Liquid - Liquid .2 Liquid - Solid .3 Solid - Solid	Alternating Running	2 - 4	1- Catastrophic a. Fails While Running b. Seal Rupture - External c. Seal Rupture - Influx d. Spurious Start/Command Fault e. Fails to Start on Demand f. Fails to Stop on Demand
3.3.6.2 <u>Blenders, Solid - Solid</u>				

3.0 PROCESS EQUIPMENT
3.3 Rotating Equipment

APPENDIX A - CCPS TAXONOMY

EQUIPMENT DESCRIPTION			SERVICE DESCRIPTION		FAILURE DESCRIPTION
			OPERATING MODE	PROCESS SEVERITY	
3.3.7 Pumps 3.3.7.1 Air 3.3.7.2 Motor-Driven 3.3.7.2.1 Pressure 3.3.7.2.2 Vacuum 3.3.7.3 Turbine-Driven 3.3.7.3.1 Gas 3.3.7.3.2 Steam	 .1 Centrifugal .2 Positive Displacement .2.1 Gear .2.2 Piston .1 Centrifugal .2 Positive Displacement	.1 0 - 1000 HP .2 >1000 HP	Alternating Running Standby	1 - 4	1- Catastrophic a. Fails While Running b. Rupture c. Spurious Start d. Fails to Start on Demand e. Fails to Stop on Demand 2- Degraded a. Fails to Run at Rated Speed b. External Leak 3- Incipient a. High Vibration b. Over-temperature c. Over-current
3.3.8 Rotary Agitators 3.3.8.1 Direct-Driven 3.3.8.2 Gear-Driven			Alternating Running	2 - 4	1- Catastrophic a. Fails While Running b. Seal External Rupture c. Seal Rupture - Influx d. Spurious Start/Command Fault e. Fails to Start on Demand f. Fails to Stop on Demand

3.0 PROCESS EQUIPMENT
3.4 Solids Handling

APPENDIX A - CCPS TAXONOMY

EQUIPMENT DESCRIPTION	SERVICE DESCRIPTION		FAILURE DESCRIPTION
	OPERATING MODE	PROCESS SEVERITY	
3.4.1 Baggers/Packagers			1- Catastrophic a. Fails While Running b. Rupture (External Leakage) c. Internal Leakage d. Fails to Start on Demand e. Fails to Stop on Demand
3.4.2 Conveyors **3.4.2.1** <u>Belt</u> **3.4.2.2** <u>Screw</u>			1- Catastrophic a. Fails While Running b. Rupture (External Leakage) c. Internal Leakage d. Fails to Start on Demand e. Fails to Stop on Demand 2- Degraded
3.4.2.3 <u>Pneumatic</u>			1- Catastrophic a. Leakage 0 - 1/4" b. Leakage >1/4" c. Rupture 2- Degraded 3- Incipient a. Wall Thinning b. Embrittlement c. Cracked or Flawed
3.4.3 Elevators			1- Catastrophic a. Fails While Running b. Rupture (External Leakage) c. Internal Leakage d. Fails to Start on Demand e. Fails to Stop on Demand

APPENDIX A - CCPS TAXONOMY

EQUIPMENT DESCRIPTION	SERVICE DESCRIPTION		FAILURE DESCRIPTION
	OPERATING MODE	PROCESS SEVERITY	
3.4.4 Feeders 3.4.4.1 Rotary 3.4.4.2 Vibrating			1- Catastrophic a. Fails While Running b. Rupture (External Leakage) c. Internal Leakage d. Fails to Start on Demand e. Fails to Stop on Demand
3.4.5 Separators 3.4.5.1 Bag Houses			1- Catastrophic a. Fails While Running b. External Leakage 0 - 1/4" c. External Leakage >1/4" d. Rupture e. Internal Leakage f. Fails to Start on Demand g. Fails to Stop on Demand 2- Degraded 3- Incipient a. Wall Thinning b. Embrittlement c. Cracked or Flawed
3.4.5.2 Cyclones			1- Catastrophic a. Leakage 0 - 1/4" b. Leakage >1/4" c. Rupture 2- Degraded 3- Incipient a. Wall Thinning b. Embrittlement c. Cracked or Flawed

APPENDIX A - CCPS TAXONOMY

EQUIPMENT DESCRIPTION	SERVICE DESCRIPTION		FAILURE DESCRIPTION
	OPERATING MODE	PROCESS SEVERITY	
3.4.5 Separators (Cont'd.) 3.4.5.3 <u>Classifiers</u>			1- Catastrophic a. Fails While Running b. Rupture (External Leakage) c. Internal Leakage d. Fails to Start on Demand e. Fails to Stop on Demand
3.4.5.4 <u>Dust Collectors</u>			1- Catastrophic a. Fails While Running b. External Leakage 0 - 1/4" c. External Leakage >1/4" d. Rupture e. Internal Leakage f. Fails to Start on Demand g. Fails to Stop on Demand 2- Degraded 3- Incipient a. Wall Thinning b. Embrittlement c. Cracked or Flawed
3.4.5.5 <u>Sifters</u>			1- Catastrophic a. Fails While Running b. Rupture (External Leakage) c. Internal Leakage d. Fails to Start on Demand e. Fails to Stop on Demand
3.4.5.6 <u>Screens</u> 3.4.5.6.1 Rotary 3.4.5.6.2 Vibrating			1- Catastrophic a. Fails While Running b. Rupture (External Leakage) c. Internal Leakage d. Fails to Start on Demand e. Fails to Stop on Demand

APPENDIX A - CCPS TAXONOMY

EQUIPMENT DESCRIPTION	SERVICE DESCRIPTION		FAILURE DESCRIPTION
	OPERATING MODE	PROCESS SEVERITY	
3.4.5 Separators (Cont'd.) 3.4.5.7 Magnets			1- Catastrophic a. Leakage 0 - 1/4" b. Leakage >1/4" c. Rupture 2- Degraded 3- Incipient a. Wall Thinning b. Embrittlement c. Cracked or Flawed
3.4.5.8 Electrostatic Precipitators			1- Catastrophic a. Fails While Running b. External Leakage 0 - 1/4" c. External Leakage >1/4" d. Rupture e. Internal Leakage f. Fails to Start on Demand g. Fails to Stop on Demand 2- Degraded 3- Incipient a. Wall Thinning b. Embrittlement c. Cracked or Flawed
3.4.6 Size Reducers 3.4.6.1 Crushers 3.4.6.2 Chippers 3.4.6.3 Cutters 3.4.6.4 Granulators 3.4.6.5 Grinders 3.4.6.6 Mills			1- Catastrophic a. Fails While Running b. Rupture (External Leakage) c. Internal Leakage d. Fails to Start on Demand e. Fails to Stop on Demand

3.0 **PROCESS EQUIPMENT**
3.5 **Valves**

APPENDIX A - CCPS TAXONOMY

EQUIPMENT DESCRIPTION			SERVICE DESCRIPTION		FAILURE DESCRIPTION
			OPERATING MODE	PROCESS SEVERITY	
3.5.1 **Check Valves** 3.5.1.1 Operated Stop Check 3.5.1.1.1 Manual 3.5.1.1.2 Motor 3.5.1.1.3 Pneumatic 3.5.1.2 Non-Operated Check 3.5.1.2.1 Swing Check 3.5.1.2.2 Stop Check				1 - 4	1- Catastrophic a. Fails to Check b. Fails to Open c. Fails to Re-open 2- Degraded a. Significant Back-leakage
3.5.2 **Manual Valves**	.1 Normally Open .2 Normally Closed	.1 0 - 2" .2 >2 - 12" .3 >12"		1 - 4	1- Catastrophic a. Leakage 0 -10% b. Leakage >10% c. Rupture d. Normally Open/Fails Open e. Normally Closed/Fails Closed f. Normally Open/Fails Plugged Closed g. Normally Closed/Fails Open (Internal Leak) 2- Degraded 3- Incipient a. Wall Thinning b. Embrittlement c. Cracked or Flawed d. Internal Leakage

APPENDIX A - CCPS TAXONOMY

EQUIPMENT DESCRIPTION		SERVICE DESCRIPTION		FAILURE DESCRIPTION
		OPERATING MODE	PROCESS SEVERITY	
3.5.3 **Operated Valves** 3.5.3.1 Hydraulic	.1 Fail Close Actuator .2 Fail Open Actuator .3 Fail in Position Actuator	.1 Alternating (2-position Control) .2 Continuous (Throttling Control) .3 Standby Open .4 Standby Closed	1 - 4	All Modes Catastrophic a. External Leakage b. Internal Leakage > 1% c. Spurious Operation d. No Change of Position on Demand Degraded a. Delayed Actuation Incipient a. Wall Thinning b. Embrittlement c. Cracked or Flawed d. Internal Leakage
3.5.3.2 Motor	Same as above	Same as above	1 - 4	
3.5.3.3 Pneumatic	Same as above	Same as above	1 - 4	
3.5.3.4 2-Way Solenoid	.1 Single Coil Acturator	.1 Alternating (2-Position Control) .2 Energized Standby Open .3 Energized Standby Closed .4 Deenergized Standby Open .5 Deenergized Standby Closed	1 - 4	
	.2 Dual Coil Actuator	.1 Alternating .2 Standby Open .3 Standby Closed		
3.5.3.5 3-Way Solenoid	.1 Single Coil Actuator	.1 Alternating (2-Position Control) .2 Energized .3 Deenergized	1 - 4	All Modes Catastrophic a. Internal Leakage b. Vent Port Plugged c. Spurious Operation d No Change of Position on Demand
	.2 Dual Coil Actuator	.1 Alternating .2 Standby		

APPENDIX A - CCPS TAXONOMY

EQUIPMENT DESCRIPTION					SERVICE DESCRIPTION		FAILURE DESCRIPTION
					OPERATING MODE	PROCESS SEVERITY	
3.6.1 Atmospheric **3.6.2 Pressurized** **3.6.3 Vacuum**	.1 Metallic .2 Non-Metallic	.1 Field-Fabricated .2 Shop Fabricated	.1 Internals .2 No Internals	.1 Reacting .2 Non-Reacting		1 - 4	1- Catastrophic a. Leakage >1/4" b. Leakage 0 -1/4" c. Rupture d. Plugging 2- Degraded a. Restricted Flow 3- Incipient a. Wall Thinning b. Embrittlement c. Cracked or Flawed d. Erratic Flow

APPENDIX A - CCPS TAXONOMY

EQUIPMENT DESCRIPTION					SERVICE DESCRIPTION		FAILURE DESCRIPTION
					OPERATING MODE	PROCESS SEVERITY	
3.7.1 Electrolytic Cells **3.7.2 Seals/Gaskets** 3.7.2.1 <u>Active</u>	.1 Rotating .2 Sliding	.1 Lube .2 Non-Lube	.1 Flow .2 Non-Flow .1 Single Mechanical .2 Double Mechanical	.1 0-15 psig .2 >15-300 psig .3 >300 psig		1 - 4	Active Failure Modes: 1- Catastrophic a. Seizing (implies that shaft or stem stops) b. Leakage c. Rupture 2- Degraded a. Binding
3.7.2.2 <u>Static</u> 3.7.2.2.1 Flat 3.7.2.2.2 Ring 3.7.2.2.3 Wound	.1 Metallic .2 Non-Metallic .3 Metal/ Non-Metal Composite				Alternating (>300 cycles/yr) Running (0-10 cycles/yr) Standby (10-300 cycles/ yr)		Static Failure Modes: 1- Catastrophic a. Leakage b. Blowout

4.0 PROTECTION SYSTEMS

APPENDIX A - CCPS TAXONOMY

EQUIPMENT DESCRIPTION		SERVICE DESCRIPTION		FAILURE DESCRIPTION
		OPERATING MODE	PROCESS SEVERITY	
4.1 Corrosion **4.1.1 Cathodic** **4.1.2 Chemical** **4.2 Fire** **4.2.1 Explosion Suppression Systems** **4.2.2 Fire Detection Systems** **4.2.2.1** IR **4.2.2.2** Rate of Temperature Rise **4.2.2.3** Specific Ion **4.2.2.4** UV **4.2.3 Fire Suppression Systems** **4.2.3.1** CO_2 Systems **4.2.3.2** Water Systems **4.2.3.3** Dry Powder Systems **4.2.3.4** Halon Systems				1- Catastrophic a. Fails to Operate b. Spurious Operation c. Plugged d. Leaks Through 2- Degraded a. Improper Operation b. Minor Leak Through 3- Incipient a. External Leakage b. Faulty Indication
4.2.4 Fire Water Pumps **4.2.4.1** Diesel **4.2.4.2** Electric	.1 Gear/Shaft Driven .2 Hydraulically Driven			1- Catastrophic a. Fails to Start b. Fails while Running 2- Degraded a. Low Output 3- Incipient a. Vibration b. Leakage
4.2.5 Atmospheric Gas Detection Systems **4.2.5.1** Flammable **4.2.5.2** Toxic				1- Catastrophic a. Failure to Operate on Demand b. Spurious Operation

APPENDIX A - CCPS TAXONOMY

EQUIPMENT DESCRIPTION		SERVICE DESCRIPTION		FAILURE DESCRIPTION
		OPERATING MODE	PROCESS SEVERITY	
4.3 <u>Pressure</u> **4.3.1 Blow-down Valves** **4.3.1.1** <u>Hydraulically Operated</u> **4.3.1.2** <u>Pneumatically Operated</u>	.1 Hydrocarbon, Oil .2 Hydrocarbon, Gas		2 - 4	1- Catastrophic a. Fails to Operate b. Spurious Operation c. Plugged d. Leaks Through 2- Degraded a. Improper Operation b. Minor Leak Through 3- Incipient a. External Leakage b. Faulty Indication
4.3.2 Bursting Discs/Diaphragms **4.3.3 Safety Relief Valves** **4.3.3.1** <u>Pilot Operated</u> **4.3.3.2** <u>Spring-Loaded</u>	.1 Soft Seat .2 Metal Seat		1 - 4	1- Catastrophic a. Seat Leakage b. Fails to Open c. Spurious Operation c.1 Opens Prematurely c.2 Failure to Reclose once Open d. Fails to Open on Demand 2- Degraded a. Interstage Leakage 3- Incipient a. Pilot Leakage

5.0 UTILITIES

APPENDIX A - CCPS TAXONOMY

EQUIPMENT DESCRIPTION	SERVICE DESCRIPTION		FAILURE DESCRIPTION
	OPERATING MODE	PROCESS SEVERITY	
5.1 Cooling Water Systems 5.2 Flares 5.3 Gas Generators 5.3.1 Inert Gas Generation 5.3.2 Nitrogen Vaporization 5.4 Heating Systems 5.4.1 Hot Oil 5.4.2 Salt/Eutectic 5.5 HVAC 5.6 Incinerators 5.7 Refrigeration 5.8 Steam Systems 5.8.1 Fossil Fuel 5.8.2 Waste Heat			

Appendix B
Equipment Index

Sections 2.4 and Section 3.4 discuss the use of this Equipment Index to help locate a taxonomy number in the CCPS Generic Data Base.

Taxonomy No.	*Equipment*
3.6	Absorbers
3.6	Accumulators
N/A	Actuators
3.6	Adsorbers
3.3.8	Agitators
N/A	Alarm horns
N/A	Alternators
N/A	Amplifiers
2.1.1	Analyzers-process wetted/field instruments
2.2.2	Annunciators-control room
4.2.5	Atmospheric gas detector systems
3.6.1	Atmospheric tanks
N/A	Atomizers
3.4.5.1	Bag houses
3.4.1	Baggers
1.2.1	Batteries
1.2.2	Battery chargers
N/A	Bearings
N/A	Bellows
N/A	Belt drives
3.6.1	Bins
3.3.6	Blenders
3.3.6.1.1	-liquid
3.3.6	-ribbon
3.3.6	-rotating shell
3.3.6.2.1	-solid/solid
4.3.1	Blow-down valves
3.3.3	Blowers
N/A	Boards (printed circuit, terminal)
3.1	Boilers
N/A	Bolts
N/A	Bourdon tubes

Taxonomy No.	*Equipment*
N/A	Brakes
N/A	Burners
4.3.2	Bursting discs/diaphragms
N/A	Buzzers
3.1.1	Calciners-fired
N/A	Capacitors
3.3.1	Centrifuges
3.5.1	Check valves
N/A	Chimneys
3.4.6.2	Chippers
3.7.1	Chlorine cells
2.1.1.4.4	Chromatographic analyzers
1.2.3	Circuit breakers
3.6	Clarifyers
3.4.5.3	Classifiers
N/A	Coils
3.6	Columns (packed, bubble plate, spray, wetted wall)
3.3.2	Compressors
N/A	Computer peripherals
N/A	Concentrators (magnetic, spiral)
3.1.2	Condensers
3.5.3	Control valves
2.2.1	Controllers-control room
2.1.2	-process wetted/field
N/A	Converters (analog to digital, digital to analog)
3.4.2	Conveyors
3.4.2.1	-belt
3.4.2.3	-pneumatic
3.4.2.2	-screw
3.1.2	Coolers-fin fan
5.1	Cooling water systems
4.1.1	Corrosion protection systems-cathodic
4.1.2	Corrosion systems-chemical
N/A	Couplings
3.4.6.1	Crushers
3.6	Crystallizers
3.4.6.3	Cutters
3.4.5.2	Cyclones
N/A	Cylinders (gas)
3.6	Deaerators
3.6	Decanters
N/A	Defoamers
N/A	Diaphragms

Taxonomy No.	*Equipment*
3.6	Dissolving tanks
N/A	Drums
3.1.2	Dryers-agitated
3.1.2	-fluid bed
3.1	-rotary
3.1	-spray
N/A	Ducting
3.4.5.4	Dust collectors
N/A	Ejectors
3.2	Elbows
1.2.1	Electric batteries
1.1.1	Electric motors-AC
1.1.2	-DC
N/A	Electrical connections
3.7.1	Electrolytic cells
3.4.5.8	Electrostatic precipitators
3.4.3	Elevators-bucket
N/A	Eliminators-mist
1.3.1	Emergency power generators
N/A	Engines
3.1.2	Evaporators
3.1.2	-falling film
3.1.2	-scraped wall
3.1.2	-vacuum
N/A	Expansion joints
4.2.1	Explosion suppression systems
3.3.5	Extruders
3.3.4	Fans
3.4.4	Feeders
3.4.4.1	-rotary
3.4.4.2	-vibrating
3.6	Fermenters
3.6	Filters
3.6	-plate and frame
3.6	-pressure leaf
3.6	-vacuum rotary
4.2.2	Fire detection systems
4.2.2.1	-IR
4.2.2.2	-rate of temperature rise
4.2.2.3	-specific ion
4.2.2.4	-UV
4.2.3	Fire suppression systems
4.2.3.1	-CO_2
4.2.3.3	-dry powder
4.2.3.4	-Halon
4.2.3.2	-water systems

Taxonomy No.	*Equipment*
4.2.4	Fire water pumps
4.2.4.1	-diesel
4.2.4.2	-electric
3.1.1	Fired heaters
3.2.2.2	Fittings-lined pipe
3.2.1.2	-metal pipe
3.2.3.2	-rigid plastic pipe
3.2.4.2	-tubing
3.4	Flakers-rotary drum
2.1.5	Flame detectors
N/A	Flammable gas detection systems
N/A	Flanges
5.2	Flares
2.1.6.1	Flow meters
N/A	Fluidized beds
3.1.1	Furnaces
1.2.6	Fuses
4.2.5	Gas detection systems-atmospheric
N/A	Gas distributors
5.3	Gas generators
3.6.1	Gas holders
3.7.2	Gaskets
N/A	Gauges
N/A	Gears
N/A	Generators
N/A	Governors
3.4.6.4	Granulators
3.4.6.5	Grinders
3.1.1	Heat exchangers-fired
3.1.2	-plate
3.1.2	-shell and tube
3.1.2	-spiral
3.1.2	-unfired
3.1.2	Heaters-electric heated
3.1.1	-fossil fuel fired
3.1.2	-steam heated
5.4.1	Heating systems-hot oil
5.4.2	-salt/eutectic
3.3.6	Homogenizers
3.6.1	Hoppers
3.2.5	Hoses
5.5	HVAC
N/A	Impellers
5.6	Incinerators

Taxonomy No.	*Equipment*
2.2.3	Indicators-control room
2.1.6	-process wetted/field
N/A	Inductors
5.3.1	Inert gas generators
N/A	Injectors
2.2	Instrumentation/control room
2.2.2	-annunciators
2.2.5	-computational modules
2.2.1	-controllers
2.2.10	-distributed control systems
2.2.3	-indicators(level,flow,etc.)
2.2.7	-power supplies
2.2.11	-programmable controllers
2.2.4	-recorders
2.2.9	-switches
2.2.6	-totalizers
2.2.8	-weight scales
2.1	Instrumentation/process wetted/field
2.1.1	-analyzers
2.1.2	-controllers
2.1.5	-flame detectors
2.1.6	-indicators
2.1.7	-recorders
2.1.4	-switches
2.1.8	-transducers
2.1.3	-transmitters
N/A	Integrated circuits
N/A	Interlock circuits
1.2.4	Inverters-drive system power
N/A	Ion exchangers
N/A	Joints
3.1.1	Kilns-rotary
3.1.1	-stationary
N/A	Lamps (inside alarms)
2.1.6.2	Level indicators
N/A	Limit switches
3.4.5.7	Magnets
1.3.2	Main power generators
N/A	Manifolds
N/A	Manometers
3.5.2	Manually operated valves
N/A	Mesh, wire
N/A	Meters (e.g. flow, moving coil)
3.4.6.6	Mills

Taxonomy No.	*Equipment*
N/A	Mist eliminators
3.3.8	Mixers-agitators
3.3.6	-blenders
1.1.1	Motors-electric-AC
1.1.2	-DC
5.3.2	Nitrogen vaporizer systems
N/A	Nozzles
N/A	Nuts
3.5.3	Operated valves
N/A	Orifices
3.1	Ovens
4.3	Overpressure protection systems
3.4.1	Packagers
3.6	Packed towers
N/A	Packing
3.3.5	Pelletizers
3.6	Phase separators
N/A	Pins
	Pipe connections/connecting fittings
3.2.2.3	-lined
3.2.1.4	-metal
3.2.3.3	-rigid plastic
	Pipe fittings
3.2.2.2	-lined
3.2.1.2	-metal
3.2.3.2	-rigid plastic
	Pipe penetrations (thermowells, etc.)
3.2.2.4	-lined
3.2.1.3	-metal
3.2.3.4	-rigid plastic
3.2	Pipe systems
3.2.2	-lined
3.2.1	-metal
3.2.3	-rigid plastic
3.2.1.5	Pipe welds-metal pipe
3.2.2	Pipes-lined-all categories
3.2.1	-metal-all categories
3.2.3	-rigid plastic all-categories
3.2.2.1	Pipes-straight section-lined
3.2.1.1	-metal
3.2.3.1	-rigid plastic
N/A	Piping-ducting
N/A	Pistons
3.4.2.3	Pneumatic conveyors
N/A	Positioners

Taxonomy No.	*Equipment*
1.3.1	Power generators-emergency
1.3.2	-main
2.2.7	Power supply-control room
1.2.9	-uninterruptable
1.2.8	Power transformers
N/A	Precipitators
N/A	Presses
2.1.6.3	Pressure indicators
3.6.2	Pressure vessels
N/A	Propellers
N/A	Pulleys
N/A	Pulverizers
3.3.7	Pumps
3.3.7.1	-air operated
3.3.7	-all types and drives
3.3.7	-centrifugal, see drive type
3.3.7.2	-motor operated
3.3.7	-positive displacement, see drive type
3.3.7	-reciprocating, see drive type
3.3.7.3	-turbine driven
3.3.7.2.2	-vacuum
N/A	Purifiers
N/A	Quench tanks
3.6	Reactors
3.1	Reboilers
2.2.4	Recorders-control room
2.1.7	-process wetted/field
N/A	Rectifiers-drive system
N/A	-electrochemical processing
5.7	Refrigeration systems
N/A	Regenerators
N/A	Regulators
1.2.7	Relays-power
4.3.3	Relief valves
N/A	Resistors
N/A	Restrictors (flow)
N/A	Retaining rings
N/A	Rotary cutters
3.4.4.1	Rotary feeders
3.4.5.6.1	Rotary screens
4.3.2	Rupture discs/diaphragms
4.3.3	Safety relief valves
4.3.3.2	-spring loaded
4.3.3.1	-pilot operated

Taxonomy No.	*Equipment*
3.4.5.6	Screens
3.4.5.6.1	-rotary
3.4.5.6.2	-vibrating
3.4.2.2	Screw conveyors
N/A	Screws
3.6	Scrubbers
3.7.2	Seals
N/A	Semiconductor diodes
N/A	Sensors
3.4.5	Separators
3.6	Settlers
N/A	Shafts
N/A	Shredders
N/A	Sieves
3.4.5.5	Sifters
3.6.1	Silos
N/A	Slip ring assemblies
	Solenoid valves
3.5.3.4	-two-way
3.5.3.5	-three-way
N/A	Solid state devices
3.6.2	Spheres
N/A	Springs
N/A	Stacks
N/A	Static mixers
5.8.1	Steam systems-fossil fuel
5.8.2	-waste heat
N/A	Sterilizers
N/A	Stills
2.2.9	Switches-instruments-control room
2.1.4	-instruments-process wetted/field
N/A	-limit
N/A	Switchgear bus
N/A	Tachometers
3.6.1	Tanks-atmospheric pressure
3.6.2	-pressure
3.6.3	-vacuum
2.1.6.4	Temperature indicators
2.1.6.4.3	Thermometers
3.2.1.3	Thermowells
3.5.3.5	Three-way solenoid valves
N/A	Timers
2.2.6	Totalizers-control room
N/A	Towers
N/A	Tracing

Taxonomy No.	*Equipment*
2.1.8	Transducers
1.2.8	Transformers
N/A	Transistors
2.1.3	Transmitters-instruments
3.6	Trayed towers
N/A	Trays
3.2.4.1	Tubing
3.2.4.2	Tubing fittings
3.2.4	Tubing systems
See driven equipment	Turbines
1.2.9	Uninterruptable power supplies
3.6.3	Vacuum tanks
3.5	Valves
3.5.1	-backflow preventer
3.5.1	-check
3.5.3.1	-control-hydraulically operated
3.5.3.2	-control-motor operated
3.5.3.3	-control-pneumatically operated
3.5.2	-manually operated
N/A	-positioner
4.3.3	-safety
3.5.2	-shut off-manually operated
3.5.3	-shut off-with an operator
3.5.3.4	-solenoid operated
3.5.3.5	-solenoid operated three-way
3.5.3	-with an operator
N/A	Vanes
N/A	Vaporizers
N/A	Vents
3.6.1	Vessels-atmospheric
3.6.2	-pressure
3.6.3	-vacuum
3.4.4.2	Vibrating feeders
3.4.5.6.2	Vibrating screens
N/A	Voltage regulation
N/A	Weight measurement
2.2.8	Weight scales-control room
N/A	Weirs
3.2.1.5	Welds
N/A	Wire mesh
N/A	Wires and wiring

Appendix C
Matrix of Data Elements in Data Resources

The purpose of this appendix is to provide the reader with additional information on the nature of the failure rate data presented by the data resources summarized in Chapter 4. Each of these resources has been examined to determine which of the following data elements are used to present and qualify the reliability data:

Equipment type: Describes the hardware to which failure rate data apply, for example, pumps and valves. (See Appendix B.)

Component type: Describes the hardware parts to which the failure rate data apply; a further breakdown of *equipment type*.

Failure severity: The degree of functional degradation of hardware usually noted through deficient performance; categorized as "catastrophic," "degraded," and "incipient."

Failure mode: A symptom, condition, or fashion in which hardware fails. A mode might be identified as loss of function; premature function (function without demand); an out of tolerance condition; or a simple physical characteristic such as a leak (incipient failure mode) observed during inspection.

Operating mode: The method of operating equipment; classified as "alternating," "running," or "standby." (See Glossary.)

Number of failures: The number of time-related or demand-related events that have been recorded for a given *exposure time* or *number of demands*.

Exposure time: The historical operating time of the equipment population.

Number of demands: The total number of demands experienced by the equipment population within the time frame established for the study.

Mean: The measure of central tendency of a distribution, often referred to as its arithmetic average.

Error bounds: The end points of that portion of a confidence interval which is expected to contain the *mean*.

In the following data tables, "X"s have been used to indicate which data elements are contained in each data resource in column 1. The resulting information can be used by the reader to help select data resources. Note that data resources may contain more data elements than are indicated.

4.3 PROCESS EQUIPMENT DATA BASES

NO.	TITLE	INDUSTRY	Geog	Equipment Type	Component Type	Failure Severity	Failure Mode	Operating Mode	Number of Failures	Number of Demands	Exposure Time	Mean	Error Bounds
4.3-1	Imperial Chemical Industries Reliability Data Book	Chemical Process	U.K.	Misc								X	
4.3-2	Development of an Improved Liquified Natural Gas Plant Failure Rate Data Base	Chemical Process	U.S.	Pumps, Vessels, Utilities		X			X		X		
4.3-3	COMPI: Data Bank for Component Failure Data	Varied	Neth.	Mechanical, Electrical, Electronic			X		X		X	X	
4.3-4	Fluor Daniel Inc.	Chemical Process, Power, Petrochemical, Telecommunications, Nuclear Fuel Cycle	Varied	Varied								X	
4.3-5	Computerized Library of Equipment Failures (CLEF)	Chemical Process, Petroleum, Nuclear	Varied	Major equipment items found in the CPI		X	X	X	X	X		X	X
4.3-6	HARIS - Hazards and Reliability Information System-Reliability Data Base	Chemical Process, Petroleum, Natural Gas, and Nuclear	Varied	Pumps, Compressors, Gas Turbines, Valves, Vessels, Heat Exchangers, etc.			X					X	

4.4 PROCESS EQUIPMENT DATA SOURCES

NO.	TITLE	INDUSTRY	Geog	Equipment Type	Component Type	Failure Severity	Failure Mode	Operating Mode	Number of Failures	Number of Demands	Exposure Time	Mean	Error Bounds
4.4-1	Pressure Vessel Reliability	Chemical Process & Power	U.S. and Foreign	Power Boilers, Press.Vessels		X	X		X		X	X	X
4.4-2	Safety of Interstate Natural Gas Pipelines	Power	U.S.	Pipelines			X		X		X		
4.4-3	Some Data on the Reliability of Pressure Equipment in the Chemical Plant Environment	Chemical Process	U.S.and U.K.	Pressure Vessels, Heat exchangers	X				X		X	X	
4.4-4	Some Data on the Reliability of Instruments in the Chemical Plant Environment	Chemical Process	U.S. and U.K.	Instruments	X				X		X	X	
4.4-5	Failure and Maintenance Data Analysis at a Petrochemical Plant	Chemical Process	U.K.	Pumps, Compressors Conveyors, Others	X		X		X		X		
4.4-6	Causes of Ammonia Plant Shutdowns: Survey V	Chemical Process	Varied	Ammonia Plant				X					
4.4-7	Reliability Data Collection and Use in Risk and Availability Assessment	Varied	U.S. and Europe	Wide Variety	X		X	X					
4.4-8	Pipeline Reliability: An Investigation of Pipeline Failure Characteristics and Analysis of Pipeline Failure Rates	Petroleum and Natural Gas	U.S., North Sea	Pipelines	X				X		X		
4.4-9	Fault Tree Analysis Report for Coal-Gasification Process-Development Unit	Coal Gasification	U.S.	Coal unit systems			X					X	
4.4-10	Data Base Development and Equipment Reliability for Phase 1 of the Probabilistic Risk Analysis DPST-87-642	Chemical Process, Nuclear Fuel Cycle	U.S.	Pumps, fans,	X		X		X	X	X	X	X

4.4 PROCESS EQUIPMENT DATA SOURCES (CONT'D.)

NO.	TITLE	INDUSTRY	Geog	Equipment Type	Component Type	Failure Severity	Failure Mode	Operating Mode	Number of Failures	Number of Demands	Exposure Time	Mean	Error Bounds
4.4-11	Reliability Analysis of Pumps for Uranium Solutions	Chemical Process, Nuclear	U.S.	Pumps	X		X	X			X	X	
4.4-12	Emergency Generators: A Reliability Study Based on an Analysis of Failures	Chemical Process	U.S.	Emergency Diesel Generators			X	X	X		X		
4.4-13	Reliability of a Solids-Fluid Handling Process Plant	Ore Processing		Solids Handling Equipment	X		X		X			MTBF	
4.4-14	Reliability Assessment of Safety/Relief Valves	Chemical Process		Safety/Relief Valves			X		X		X		

4.5 CHEMICAL PROCESS QUANTITATIVE RISK ASSESSMENTS

NO.	TITLE	INDUSTRY	Geog	Equipment Type	Component Type	Failure Severity	Failure Mode	Operating Mode	Number of Failures	Number of Demands	Exposure Time	Mean	Error Bounds
4.5-1	Hazardous Waste Tank Failure	Chemical Process	U.S.	Vessels	X		X					X	
4.5-2	Risk Analysis of Six Potentially Hazardous Industrial Objects in the Rijnmond Area; A Pilot Study	Chemical Process	Neth.	Pumps, valves, pipe, vessels, instruments	X	X	X	X	X	X	X	X	X
4.5-3	CANVEY: An Investigation of Potential Hazardous from Operations in the Canvey Island/Thurrock Area	Chemical Process	U.K.	N/A- Event data only	NOT APPLICABLE - CONTAINS ESTIMATES OF RISK TO PUBLIC								

4.6 NON-PROCESS EQUIPMENT DATA BASES

NO.	TITLE	INDUSTRY	Geog	Equipment Type	Component Type	Failure Severity	Failure Mode	Operating Mode	Number of Failures	Number of Demands	Exposure Time	Mean	Error Bounds
4.6-1	Centralized Reliability Data Organization Input Guide	Nuclear	U.S., Japan	Pumps, Valves, etc	X		X		X	X	X		
4.6-2	Nuclear Plant Reliability Data System	Nuclear	U.S.	Varied	X	X	X						
4.6-3	The European Reliability Data System: An Organized Information Exchange on the Operation of European Nuclear Reactors	Nuclear	Europe	Varied					X	X	X		
4.6-4	Determination of Reliability Characteristic Factors in the Nuclear Power Plant Biblis b, Gesellschaft fur Reaktorsicherheit mbH	Nuclear	W.Germany	Pumps, Valves Motors Drives	X		X					X	X
4.6-5	Generating Availiability Data System	Power	U.S.	Power Plants				UNIT AVAILABILITY DATA					
4.6-6	Reliability Data Book for Components in Swedish Nuclear Power Plants	Power	Sweden	Varied	X	X	X	X	X	X	X	X	X
4.6-7	Failure and Inventory Reporting System	Offshore oil and Gas	U.S.	Safety Valves					X		X		
4.6-8	Government Industry Data Exchange Program (GIDEP)	Varied	U.S.	Varied	X		X	X	X		X		
4.6-9	System Reliability Service	Varied	U.K.	Mechanical Instruments Electrical Electronic				CONTENTS NOT AVAILABLE FOR REVIEW					
4.6-10	SAIC Data Base	Nuclear	U.S., Europe	Pumps, Valves, Diesels, Heat Exchangers, Batteries, etc.	X	X	X	X	X	X	X	X	X

4.6 NON-PROCESS EQUIPMENT DATA BASES (CONT'D.)

NO.	TITLE	INDUSTRY	Geog	Equipment Type	Component Type	Failure Severity	Failure Mode	Operating Mode	Number of Failures	Number of Demands	Exposure Time	Mean	Error Bounds
4.6-11	The In-Plant Reliability Data Base for Nuclear Plant Components	Nuclear	U.S.	Pumps, Valves, Major Electrical	X	X	X	X	X	X	X	X	X
4.6-12	IEEE Standard 500-1984	Nuclear	U.S.	Varied	X	X	X	X	X	X	X	X	X
4.6-13	Generic Data Base for Data and Models Chapter of the National Reliability Evaluation Program Guide	Nuclear	U.S.	Varied	X		X					X	X
4.6-14	Offshore Reliability Data Handbook	Offshore Oil	Varied; North Sea	Varied	X	X	X	X	X	X	X	X	X
4.6-15	RADC Non-Electronic Reliability Notebook	Commercial, Military	U.S.	Nonelectronic Varied	X			X				X	X
4.6-16	Reliability Prediction of Electronic Equipment (Military Handbook 217E)	Government and Military	U.S.	Electronic	X			X				X	

4.7 NON-PROCESS EQUIPMENT DATA SOURCES

NO.	TITLE	INDUSTRY	Geog	Equipment Type	Component Type	Failure Severity	Failure Mode	Operating Mode	Number of Failures	Number of Demands	Exposure Time	Mean	Error Bounds
4.7-1	An Aging Failure Survey of LWR Safety Systems and Components	Nuclear	U.S.	LWR Safety System Components	X	X			X				
4.7-2	Analysis of Dependent Failure Events and Failure Events Caused by Harsh Environment Conditions	Nuclear	U.S.	26 groups of mechanical electrical	X		X		X				
4.7-3	Emergency Diesel Generator Operating Experience	Nuclear	U.S.	Diesels			X		X	X			
4.7-4	A Reviews of Issues to Improving Nuclear Power Plant Diesel Generator Reliability	Nuclear	U.S.	Diesels			X		X	X			
4.7-5	Evaluation of Diesel Unavailability and Risk Effective Surveillance Test Intervals	Nuclear	U.S.	Diesels			X		X	X			
4.7-6	Operating Experience and Aging--Seismic Assessment of Electric Motor	Nuclear	U.S.	Motors		X	X		X				
4.7-7	A Review of Emergency Diesel Generator Performance at Nuclear Power Plants	Nuclear	U.S.	Diesels			X		X	X			
4.7-8	Data Summaries of Licensee Event Reports at U.S. Commercial Nuclear Power Plants (Various Components)	Nuclear	U.S.	Pumps, Valves, Major Electrical	X		X	X				X	X
4.7-9	Pipe Break Frequency Estimation for Nuclear Power Plants	Nuclear	U.S.	Pipes		X	X	X	X		X	X	X
4.7-10	ATWS: A Reappraisal, Part 3: Frequency of Unanticipated Transients	Nuclear	U.S.	Pump Seals			X		X		X		

4.7 NON-PROCESS EQUIPMENT DATA SOURCES (CONT'D.)

NO.	TITLE	INDUSTRY	Geog	Equipment Type	Component Type	Failure Severity	Failure Mode	Operating Mode	Number of Failures	Number of Demands	Exposure Time	Mean	Error Bounds
4.7-11	Survey and Evaluation of System Interaction Events and Sources	Nuclear	U.S.	Snubbers			X		X		X		
4.7-12	A Statistical Analysis of Nuclear Power Plant (Pump and Valve) Failure Rate Variability: Some Preliminary Results	Nuclear	U.S.	Pumps, Valves	X	X	X	X	X				
4.7-13	Investigation of Valve Failure Problems in LWR Power Plants	Nuclear	U.S.	Valves	X		X		X		X	X	
4.7-14	The Reliability of Emergency Diesel Generators at U.S. Nuclear Power Plants	Nuclear	U.S.	Diesels			X		X				
4.7-15	EPRI Guide	Power	U.S.	N/A				INDEX OF VARIOUS REPORTS BY EPRI					
4.7-16	Component Failure and Repair Data for Gasification-Combined Cycle Power Generation Units	Power	U.S.	Varied	X		X					X	
4.7-17	Performance of Pipework in the British Sector of the North Sea	Offshore Oil and Gas	U.K.	Pipe, Valves, Traps, Couplings			X		X		X	X	
4.7-18	Reliability Analysis Center Handbooks	Government and Military	U.S.	Varied	X			X	X		X	X	X
4.7-19	An Analysis of Reportable Incidents for Natural Gas Transmission and Gathering Lines 1970 through June 1984	Natural Gas	U.S.	Pipe					X		X		
4.7-20	Severities of Transportation Accidents	Automotive, Airline, Truck	U.S.	Rail, Trucks, Air		X			X				
4.7-21	Pressure Vessel Failure Statistics and Probabilities	Nuclear	Varied	Pressure Vessels		X	X		X		X	X	X

4.7 NON-PROCESS EQUIPMENT DATA SOURCES (CONT'D.)

NO.	TITLE	INDUSTRY	Geog	Equipment Type	Component Type	Failure Severity	Failure Mode	Operating Mode	Number of Failures	Number of Demands	Exposure Time	Mean	Error Bounds
4.7-22	Characteristics of Pipe System Failures in Light Water Reactors	Nuclear	U.S.	Pipes	X	X	X		X		X		
4.7-23	Reliability of Emergency AC Power Systems	Nuclear	U.S.	Diesels, etc.	X	X	X		X	X			

4.8 PROBABILISTIC RISK ASSESSMENTS

NO.	TITLE	INDUSTRY	Geog	Equipment Type	Component Type	Failure Severity	Failure Mode	Operating Mode	Number of Failures	Number of Demands	Exposure Time	Mean	Error Bounds
4.8-1	Big Rock Point Probabilistic Risk Assessment	Nuclear	U.S.	Varied	X		X		X	X	X	X	X
4.8-2	Connecticut Yankee Probabilistic Safety Study	Nuclear	U.S.	Varied	X		X		X	X	X	X	X
4.8-3	Indian Point Units 2 and 3	Nuclear	U.S.	Varied	X		X		X	X	X	X	X
4.8-4	Probabilistic Risk Assessment, Limerick Generating Station	Nuclear	U.S.	Varied	X		X		X	X	X	X	X
4.8-5	Interim Reliability Evaluation Program: Analysis of the Millstone Point 1 Nuclear Power Plant	Nuclear	U.S.	Varied	X		X		X	X	X	X	X
4.8-6	Oconee-3 PRA A Probabilistic Risk Assessment of Oconee Unit 3	Nuclear	U.S.	Varied	X		X		X	X	X	X	X
4.8-7	Yankee Nuclear Power Station Probabilistic Safety Study	Nuclear	U.S.	Varied	X		X		X	X	X	X	X
4.8-8	Zion Probabilistic Safety Study	Nuclear	U.S.	Varied	X		X		X	X	X	X	X
4.8-9	Reactor Safety Study: An Assessment of Accident Risk in U.S. Commercial Nuclear Power Plants (WASH-1400)	Nuclear	U.S.	Varied	X		X					X	X

Appendix D
Unreviewed Data Bases, Data Sources, and Studies

Data Resources Uncovered too Late for Review

ASEA-ATOM Data System for Nuclear Power Plants. Box 53, 72104, Wasteras 1, Sweden.

CNET, Reliability Data Bank on Electronic Equipment. 22300, Lanniou, France.

Electronic Components Data Bank. European Space Agency, Via G. Galilei, 00044, Frascati, Italy.

Edison Electric Institute Data Bank. Edison Electric Institute, 90 Park Ave., New York, NY 10016.

ENEL Data System for Power Stations. ENEL CRTN, Bastioni di Porta Volta 10, 20121, Milano, Italy.

Failure Data Handbook for Nuclear Facilities. LNEC-Memo-69-7, from NTIS

IEN "Galileo Ferraris" Reliability Data Bank, Corso Massimo D'Azeglio, 10125, Torino, Italy.

Mankamo, T. "Optimizing Test Intervals of Stand-by Diesel Generators." In E. Lauger and J. Molloft (eds.), *Reliability in Electric and Electronic Components and Systems*. North-Holland, Amsterdam 1982.

Military Electronic Laboratory (FTL/FOA) Data Bank. Fack 10450, Stockholm 80, Sweden.

Proceedings of the Second Seminar on Reliability Data Banks, Stockholm, March 1977. FTL/MEL, FOA 3, 104509 Stockholm 80, Sweden.

Reliability Data Bank, TEMA (ENI Group). TEMA S.p.a., Via Medici del Vascello, 26-20138, Milan, Italy.

Reliability Data System for Nuclear Power Plants. SEPTEN, EDF-GDF, Cedex 8, F-92080, Paris-la-Defense, France.

Scarrone, M. and N. Piccinini. "A Reliability Data Bank for the Natural Gas Distribution Industry." *Reliability Data Collection and Use in Risk and Availability Assessments, EuReDatA Conference, March 1989*. Springer-Verlag, New York, 1989.